OVERVIEW OF MONGOLIA'S WATER RESOURCES SYSTEM AND MANAGEMENT

A COUNTRY WATER SECURITY ASSESSMENT

JULY 2020

ASIAN DEVELOPMENT BANK

ADB

© 2020 Asian Development Bank
6 ADB Avenue, Mandaluyong City, 1550 Metro Manila, Philippines
Tel +63 2 8632 4444; Fax +63 2 8636 2444
www.adb.org

Some rights reserved. Published in 2020.

ISBN 978-92-9262-285-5 (print), 978-92-9262-286-2 (electronic), 978-92-9262-287-9 (ebook)
Publication Stock No. TCS200202-2
DOI: http://dx.doi.org/10.22617/TCS200202-2

The views expressed in this publication are those of the authors and do not necessarily reflect the views and policies of the Asian Development Bank (ADB) or its Board of Governors or the governments they represent.

ADB does not guarantee the accuracy of the data included in this publication and accepts no responsibility for any consequence of their use. The mention of specific companies or products of manufacturers does not imply that they are endorsed or recommended by ADB in preference to others of a similar nature that are not mentioned.

By making any designation of or reference to a particular territory or geographic area, or by using the term "country" in this document, ADB does not intend to make any judgments as to the legal or other status of any territory or area.

Corrigenda to ADB publications may be found at http://www.adb.org/publications/corrigenda.

Note:
In this publication, "$" refers to United States dollars.

On the cover: Addressing the water security challenges in Mongolia requires an integrated approach to water resources management (photos by ADB).

Cover design by Rommel Marilla.

Contents

Tables, Figures, Boxes, and Map   v

Acknowledgments   vi

Abbreviations   vii

Glossary   viii

Executive Summary   ix

1   Introduction   1

2   The Water Resources System in Mongolia and the Key Dimensions of Water Security   3
     2.1   Natural Resources System   3
     2.2   Socioeconomic System   6
     2.3   Administrative and Institutional System   8

3   Water Management in Mongolia: History, Challenges, Trends, and Future   11
     3.1   Implementing Integrated Water Resources Management   11
     3.2   Water Challenges and Hot Spots   15

4   Water Security Assessment of Mongolia   20
     4.1   Methodology   20
     4.2   Water Security at the National Level   23
     4.3   Water Security at the River Basin Level   25

5   Institutional Assessment   29
     5.1   Key Institutional and Policy Issues for Water Security   29
     5.2   Benchmarking of River Basin Organizations   30
     5.3   Performance of the Service-Providing Sector   33
     5.4   Summary of the Institutional Review   34

6   Planning for Action   **36**

   6.1   Outline of the Physical Investments   36

   6.2   Strengthening Institutions   40

7   Summary of Lessons Learned and Recommendations   **43**

Appendixes

   1   Project Concept Notes for Possible Priority Investment Projects: Integrated Water Supply   46
      and Sanitation, Water Points, Pasture, and Livestock Production Systems

   2   Project Concept Notes for Possible Priority Investment Projects: Strategic Investment   51
      for Water Supply and Sanitation in *Ger* Areas of Ulaanbaatar

   3   Project Concept Notes for Possible Priority Investment Projects: Integrated Investment   55
      for Flood Protection in Urban Areas

Tables, Figures, Boxes, and Map

Tables

1	Water Security Stages	21
2	Adjustments to the Asian Water Development Outlook 2016 Methodology for Mongolia's Country Water Security Assessment	22
3	Water Security Scores and Core Interventions by Key Dimension	25
4	Water Security Scores and Indexes by River Basin	27
5	Water Security Key Dimensions and Associated Water Sector Institutions	29
6	Benchmarking Performance Indicators and Scores	31
7	Definition of Benchmarking Scores	32
8	Summary Findings of the Institutional Review	34

Figures

1	Asian Water Development Outlook 2016 Framework for Water Security	2
2	Water Withdrawal by River Basin and Sector, 2014	7
3	Total Water Withdrawal as a Percentage of Total Renewable Water Resources by River Basin	8
4	Overview of the Administrative and Institutional System for Integrated Water Resources Management in Mongolia	10
5	Integrated Water Resources Management—Integrating the Subsystems	12
6	The Three Pillars of Integrated Water Resources Management	13
7	Mongolia's National Water Security Scores by Indicators per Key Dimension	24
8	Mongolia's River Basin Water Security Scores	26
9	Average Benchmarking Scores across 21 River Basin Organizations	32
10	Cost Curves for the Main Areas of Incremental Water Demand, 2017	39

Boxes

1	Tuul River Basin Health Report Card	14
2	What is a *Dzud*?	18

Map

	Continental Basins of Mongolia	4

Acknowledgments

The authors of this publication—Mingyuan Fan, principal water resources specialist of the Environment, Natural Resources and Agriculture Division (EAER), East Asia Department (EARD), Asian Development Bank (ADB); and Eelco van Beek, ADB consultant—are grateful for the continuous support and guidance of ADB's EARD management, particularly to James Lynch, director general; M. Teresa Kho, deputy director general; Qingfeng Zhang, EAER director; Akiko Terada-Hagiwara, principal economist; and Ying Qian, advisor. Amy S. P. Leung, former director general, EARD, also provided insightful advice and comments during the early stage of report preparation. ADB's Mongolia Resident Mission, through Declan F. Magee, senior country economist, and Ongonsar Purev, senior environment officer, facilitated the country sensitivity and map clearance of this publication.

The authors would also like to recognize the efforts of the following peer reviewers for their valuable comments: Sanmugam Ahembaranathan Prathapar, senior water resources specialist of ADB's Sustainable Development and Climate Change Department (SDCC); and Jelle Beekma, senior water resources specialist (food security), Water Sector Group, SDCC. Hugh Milner, ADB consultant; Simon Costanzo, science integrator, University of Maryland Center for Environmental Science; and Purevdorj Surenkhorloo, freshwater and climate change officer, WWF Mongolia Programme Office in Ulaanbaatar, also contributed to this publication.

The authors also express their utmost appreciation to Mongolia's Ministry of Environment and Tourism, particularly to Myagmar Shar, director general, as well as to Munkhbat D and Saraa Borchuluun, project officers.

This publication is based on the *Country Water Security Assessment of Mongolia* (an output of the eponymous ADB technical assistance project), prepared by consultants from FCG ANZDEC and the Mongolia Water Forum. Special thanks to the project consultants, Adrian Young and P. Batimaa, and to the ADB project supervisor, Shahbano Tirmizi.

Joy Quitazol-Gonzalez reviewed the final draft of the report and facilitated its publishing from manuscript editing to engagement of service providers, proofreading, and final production process. The authors also acknowledge the assistance of Sophia Castillo-Plaza, Mary Dianne Rose Umayan, and Omar Shariff Belisario, EARD.

Abbreviations

ADB	–	Asian Development Bank
AWDO	–	Asian Water Development Outlook
CWSA	–	country water security assessment
IWMP	–	Integrated Water Management Plan
IWRM	–	integrated water resources management
KD	–	key dimension
km^2	–	square kilometer
m^3	–	cubic meter
O&M	–	operation and maintenance
OECD	–	Organisation for Economic Co-operation and Development
PRC	–	People's Republic of China
RBA	–	river basin authority
RBC	–	river basin council
RBO	–	river basin organization
SDG	–	Sustainable Development Goal
SDV	–	Sustainable Development Vision
TWS	–	terrestrial water storage
WSP	–	water service provider
WSS	–	water supply and sanitation

Glossary

aimag	–	a provincial administrative unit in Mongolia
dzud	–	a weather phenomenon unique in Mongolia where summer drought (July–September) is followed by severe winter (November–February), resulting in death of a large number of livestock
ger	–	a traditional tent used by nomadic herders of Mongolia; *ger* areas refer to Mongolia's traditional tent communities
gobi	–	a desert area
soum	–	a subprovincial administrative unit in Mongolia

Executive Summary

The extreme climatic conditions in Mongolia present a challenge for the water managers in the country to provide water security for the people. Being water secure involves providing sufficient water of good quality to the population and for economic activities, protecting against waterborne pollution and water-related disasters, and preserving the ecosystems. The Asian Development Bank (ADB), together with the Asia-Pacific Water Forum, has developed an analytical framework to measure water security as part of their Asian Water Development Outlook (AWDO) series, wherein water security scores and indexes of countries in Asia and the Pacific are compared. This publication describes the application of the AWDO approach to Mongolia, broadened with other analyses on the water security of the country.

The Water Resources System in Mongolia

Rainfall in Mongolia is relatively low, ranging from a yearly average of 350 millimeters in the north to 80 millimeters in the south in the Gobi Desert. In terms of water availability per capita, the average water endowment is high for the country as a whole. However, regional differences in rainfall and population result in local hot spots of water insecurity, particularly in Ulaanbaatar, where half the population lives, and in the *gobi* region, where mining companies depend on water for their operation. Because of the high seasonal variability of river flows and since rivers freeze in winter, groundwater is tapped as Mongolia's main water source for drinking and industrial water. At the same time, the valuable and fragile ecology of the country necessitates the maintenance of high environmental flow requirements. Mongolia should also prepare to adapt to climate change, as it might become a major challenge.

Assessing the Water Security in Mongolia

Beginning with the 2013 edition, the AWDO presented water security as a multidimensional approach, expressed in five key dimensions (KDs): household water security (KD1), economic water security (KD2), urban water security (KD3), environmental water security (KD4), and resilience to water-related disasters (KD5). The scores of each KD are determined based on data on several indicators, with a maximum score of 20. Some adjustments were made on the AWDO 2016 methodology and indicators before applying them to Mongolia's water security assessment. These adjustments, such as exclusively focusing on rural household water security (KD1), were done to reflect the specific conditions of Mongolia. Likewise, to account for regional differences within Mongolia, the adjusted country water security assessment was applied at the river basin level. The average results for the country and recommendations for interventions to improve water security in Mongolia are as follows:

(i) KD1: Rural household water security (score of 11.5)
 – Enhanced awareness and access to improved water supply and sanitation for rural herder communities.

(ii)   KD2: Economic water security (score of 12.3)

- Agriculture: expanded irrigation and livestock water points.
- Energy: increase in renewable energy by 30%; investment in water-saving measures for coal power plants.
- Mining and industry: improved water sources planning and management; increased effluent treatment.

(iii) KD3: Urban water security (score of 11.6)

- Increased options and provision for financing.
- Reduced inequities in the level of service between urban centers and *ger* areas.
- Improved levels of cost recovery for operation and maintenance to promote sustainability.

(iv)   KD4: Environmental water security (score of 16.0)

- Strengthened institutional framework for management and regulation.
- Guaranteed sustainable river and groundwater abstractions, including improved monitoring and control.

(v)   KD5: Resilience to water-related disasters (score of 14.1)

- Improved risk analysis of water-related disasters—including drought, flood, and *dzud*—to increase the cost-effectiveness of interventions.

Institutional Assessment

Increasing water security requires efficient national and regional water institutions. River basin organizations (RBOs) have the key responsibility to implement effective regional interventions. RBOs are relatively young organizations in Mongolia. Based on the 2012 Water Law, 21 RBOs (out of a planned 29) have been established, with each RBO comprising two components: a river basin authority and a river basin council. A 2016 baseline benchmarking based on a self-assessment process using 14 performance indicators revealed that the RBOs are reasonably in line with their (starting) stage of development. Main issues noted are the political will and commitment among all levels of government in relation to the role of the RBOs, adequate financial stability and sustainable cost recovery, and the need for comprehensive monitoring and evaluation.

Planning for Action

Based on the water security assessment, the report developed an investment program that sets out the details and direction of key water sector investments during 2018–2030. This action plan includes the following:

- Technical assistance for planning, stakeholder engagement, capacity building, improvement of knowledge on water resources, and strengthening of private sector partnerships.
- Strengthening of institutions, including supporting the RBOs in implementing integrated water resources management, and strengthening the water service providers in facilitating the operations of government-owned water investments.
- Adjustment of the legal and policy framework, such as simplification of the multitude of regulations, and provision of guidance on how coordination between different sectors can be improved.
- Implementation of actual investment programs related to the five KDs.

1 Introduction

Compared to many other countries, Mongolia's water resources are limited. The average annual rainfall ranges from 80 millimeters in the *gobi* region in the south to about 350 millimeters in the northern mountain areas. Seasonal variation in climatic conditions is strong, with 257 cloudless days in a year and with temperature ranging from –40 degrees Celsius (°C) in winter (November–February) to 35°C in summer (July–September). The major challenge for water resources managers is to provide the services the population expects in terms of drinking water supply; support of economic activities; and environmental protection against droughts, floods, and water pollution.

In 2018, Mongolia's population was estimated at 3.2 million people spread over a land area of 1.56 million square kilometers (km^2).[1] Mongolia is the 19th largest country in the world, yet its population density stands at only around two persons per km^2, making it the most sparsely populated fully sovereign country. Although rainfall is limited, the annual (renewable) water availability per capita is more than 10,000 cubic meters (m^3), which is high compared to most other countries. However, Mongolia's population is unevenly distributed across the country. Its capital and largest city, Ulaanbaatar, is home to about 47% of its total population. The concentration of population in urban areas puts heavy pressure on the local water resources and threatens to constrain economic development in key sectors.

The main objective of water management is to provide water security to the population. United Nations Water defines water security as "the capacity of a population to safeguard sustainable access to adequate quantities of acceptable quality water for sustaining livelihoods, human well-being, and socioeconomic development, for ensuring protection against waterborne pollution and water-related disasters, and for preserving ecosystems in a climate of peace and political stability."[2] The Asian Development Bank (ADB), together with the Asia-Pacific Water Forum, has developed an analytical framework to measure water security as part of their Asian Water Development Outlook (AWDO) series.[3] This analytical framework distinguishes five key dimensions (KDs) based on the vision that societies can enjoy water security when water resources and services are effectively managed. The KDs are as follows:

(i)   household water security (KD1)—to satisfy water supply and sanitation needs of households;

(ii)   economic water security (KD2)—to support productive economies in agriculture, energy, and industry;

(iii)   urban water security (KD3)—to develop vibrant, livable cities and towns;

(iv)   environmental water security (KD4)—to restore the health of rivers and ecosystems; and

(v)   resilience to water-related disasters (KD5)—to build resilient communities that can adapt to change.

---

[1]   Government of Mongolia, National Statistics Office. Mongolian Statistical Information Service (accessed 8 April 2019).
[2]   United Nations Water. 2013. *Water Security and the Global Water Agenda: A UN-Water Analytical Brief*. Hamilton, Ontario: United Nations University. p. 1.
[3]   ADB. 2007. *Asian Water Development Outlook 2007: Achieving Water Security for Asia*. Manila; ADB. 2013. *Asian Water Development Outlook 2013: Measuring Water Security in Asia and the Pacific*. Manila; and ADB. 2016. *Asian Water Development Outlook 2016: Strengthening Water Security in Asia and the Pacific*. Manila.

Each KD is composed of several indicators as presented in Figure 1. The score of each KD is determined by measuring the performance of these indicators.

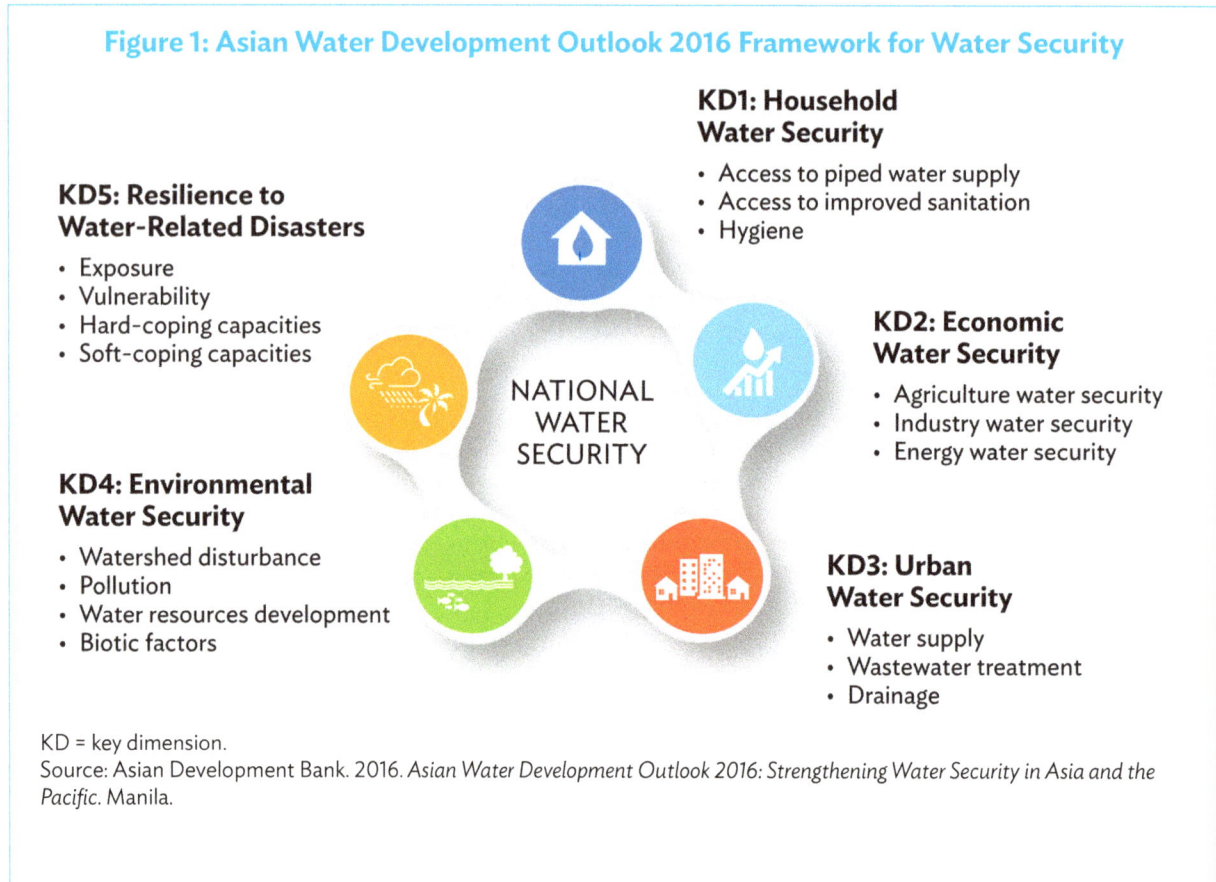

Figure 1: Asian Water Development Outlook 2016 Framework for Water Security

KD1: Household Water Security
- Access to piped water supply
- Access to improved sanitation
- Hygiene

KD5: Resilience to Water-Related Disasters
- Exposure
- Vulnerability
- Hard-coping capacities
- Soft-coping capacities

KD2: Economic Water Security
- Agriculture water security
- Industry water security
- Energy water security

NATIONAL WATER SECURITY

KD4: Environmental Water Security
- Watershed disturbance
- Pollution
- Water resources development
- Biotic factors

KD3: Urban Water Security
- Water supply
- Wastewater treatment
- Drainage

KD = key dimension.
Source: Asian Development Bank. 2016. *Asian Water Development Outlook 2016: Strengthening Water Security in Asia and the Pacific*. Manila.

This report assesses the water security in Mongolia, following the AWDO approach, particularly the methodology and indicators used in the AWDO 2016 edition, but with some adjustments to reflect the specific circumstances in Mongolia. The content is based on the country water security assessment (CWSA) study of 2017,[4] broadened with analyses from other documents and studies on the water resources situation in Mongolia—e.g., the 2013 Integrated Water Management Plan (IWMP) of Mongolia,[5] the two hydro-economic analysis reports carried out by the 2030 Water Resources Group in 2016,[6] the 2019 Tuul River Basin Health Report Card,[7] the Climate Risk Country Profile of Mongolia (2019),[8] the Intended Nationally Determined Contribution of Mongolia,[9] and the Third National Communication of Mongolia under the United Nations Framework Convention on Climate Change (2018).[10]

4 ADB. 2017. *Country Water Security Assessment of Mongolia*. Consultant's report. Manila (TA 8855-MON).
5 Government of Mongolia, Ministry of Environment and Green Development. 2013. *Integrated Water Management Plan of Mongolia*. Ulaanbaatar.
6 2030 Water Resources Group. 2016. *Hydro-Economic Analysis on Cost-Effective Solutions to Close Ulaanbaatar's Future Water Gap*. Washington, DC; and 2030 Water Resources Group. 2016. *Prioritized Solutions to Close the Water Gap: Hydro-Economic Analysis on the Coal Mining Regions in Mongolia's Gobi Desert*. Washington, DC.
7 WWF Mongolia Office; Government of Mongolia, Ministry of Environment and Tourism; Tuul River Basin Authority; University of Maryland Center for Environmental Science; and ADB. 2019. *Tuul River Basin Health Report Card 2019*. Washington, DC.
8 World Bank and ADB. 2019. *Climate Risk Country Profile: Mongolia*. Washington, DC.
9 Government of Mongolia; Ministry of Environment, Green Development and Tourism; Nature Conservation Fund; Climate Change Project Implementing Unit. 2015. *Intended Nationally Determined Contribution of Mongolia to the 2015 Agreement under the United Nations Framework Convention on Climate Change*. Ulaanbaatar.
10 Government of Mongolia, Ministry of Environment and Tourism. 2018. *Third National Communication of Mongolia under the United Nations Framework Convention on Climate Change*. Ulaanbaatar.

2 The Water Resources System in Mongolia and the Key Dimensions of Water Security

M ongolia's water resources system consists of three components: (i) the natural resources system (e.g., rivers, lakes, groundwater), which reflects the environmental and resilience aspects of water security (KD4 and KD5); (ii) the socioeconomic system (i.e., the use of water for drinking, irrigation, manufacturing, and other industries), which addresses the household, economic, and urban aspects of water security (KD1, KD2, and KD3); and (iii) the administrative and institutional system (i.e., government institutes, laws, and the private sector), which addresses the governance and financing aspects of water security.

2.1 Natural Resources System

Mongolia is located on the divide of three continental basins (as illustrated in the map on page 4): the Northern Arctic Ocean Basin, the Pacific Ocean Basin, and the Central Asian Internal Basin. The Northern Arctic Ocean Basin drains about 50% of the total river runoff of Mongolia, while the Pacific Ocean Basin drains about 10% and the Central Asian Internal Basin about 40%. These three continental basins are further divided into 29 river basins: 11 in the Northern Arctic Ocean Basin (river basins 1–11), 4 in the Pacific Ocean Basin (river basins 12–15), and 14 in the Central Asian Internal Basin (river basins 16–29).

The yearly renewable water availability is highest in the Northern Arctic Ocean Basin at 53,100 m^3 per km^2, as compared with 19,200 m^3 per km^2 in the Pacific Ocean Basin and 13,000 m^3 per km^2 in the Central Asian Internal Basin. However, since the population is also concentrated in the Northern Arctic Ocean Basin, the annual renewable water availability per capita is lowest in that basin at 8,800 m^3 per capita, as compared with the Pacific Ocean Basin's 19,000 m^3 per capita and the Central Asian Internal Basin's 16,500 m^3 per capita.

Draining to the north is the Northern Arctic Ocean Basin, which feeds the rivers of the Russian Federation that debouch into the Arctic Ocean. The main river is the Selenge River, which contributes to the large Baikal Lake in the Russian Federation. Important tributaries are the Orkhon and the Tuul rivers. The Tuul River is a tributary of the Orkhon River, while the Orkhon River joins the Selenge River just before the border with the Russian Federation. About 65% of the Mongolian population lives in the Northern Arctic Ocean Basin, where the majority of the country's socioeconomic activities occur.

The Pacific Ocean Basin drains to the east, forming the headwaters of the Amur River. This river runs along the border between the Russian Federation and Manchuria in the People's Republic of China (PRC) before debouching into the Pacific Ocean. The main river is the Onon River, which rises in the Khentii Mountain Range in northeastern Mongolia.

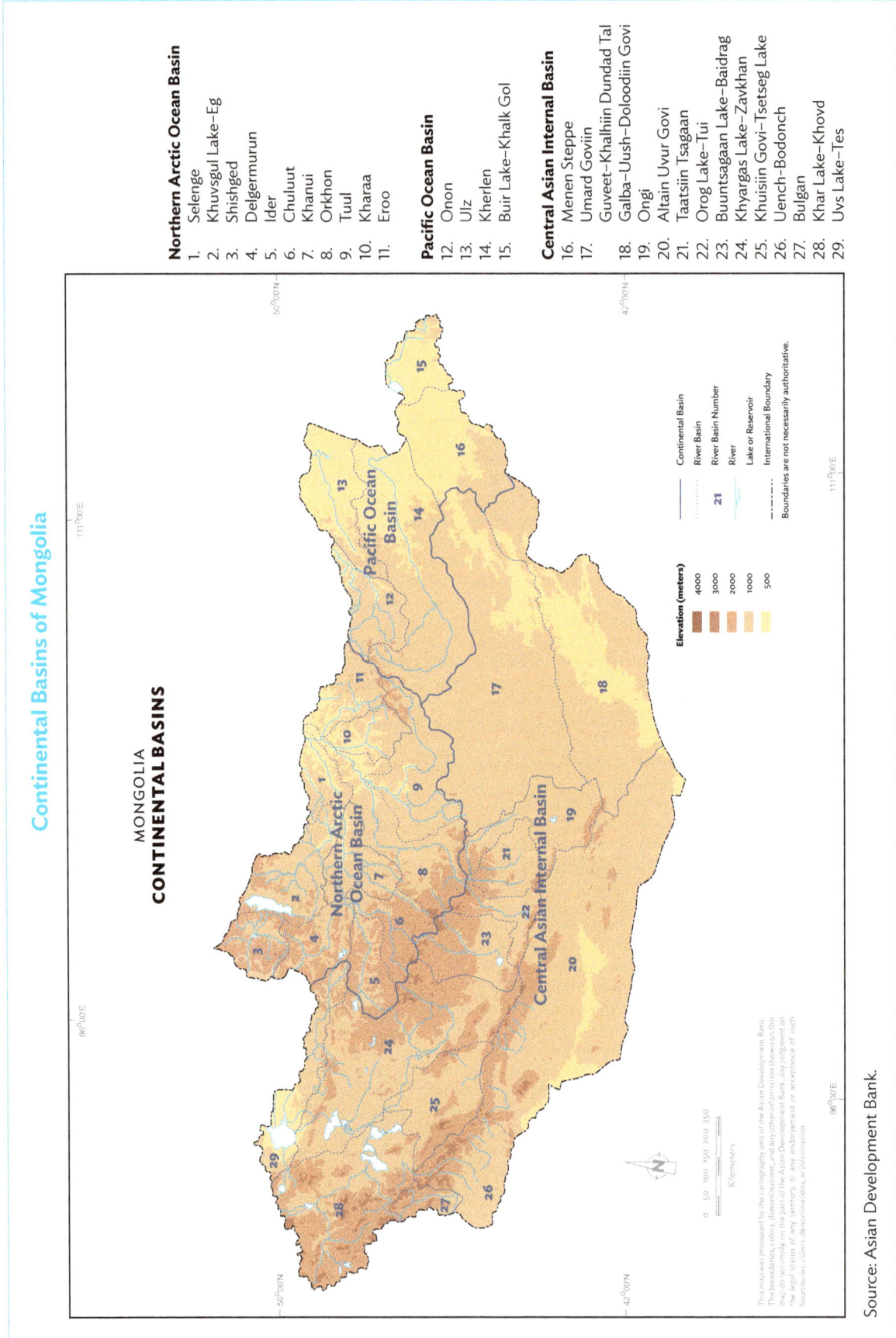

Continental Basins of Mongolia

MONGOLIA
CONTINENTAL BASINS

Northern Arctic Ocean Basin
1.   Selenge
2.   Khuvsgul Lake–Eg
3.   Shishged
4.   Delgermurun
5.   Ider
6.   Chuluut
7.   Khanui
8.   Orkhon
9.   Tuul
10. Kharaa
11. Eroo

Pacific Ocean Basin
12. Onon
13. Ulz
14. Kherlen
15. Buir Lake–Khalk Gol

Central Asian Internal Basin
16. Menen Steppe
17. Umard Goviin
18. Guveet–Khalhiin Dundad Tal
     Galba–Uush–Doloodiin Govi
19. Ongi
20. Altain Uvur Govi
21. Taatsiin Tsagaan
22. Orog Lake–Tui
23. Buuntsagaan Lake–Baidrag
24. Khyargas Lake–Zavkhan
25. Khuisiin Govi–Tsetseg Lake
26. Uench–Bodonch
27. Bulgan
28. Khar Lake–Khovd
29. Uvs Lake–Tes

Northern Arctic Ocean Basin

Pacific Ocean Basin

Central Asian Internal Basin

Elevation (meters)
4000
3000
2000
1000
500

————— Continental Basin
········· River Basin
21   River Basin Number
           River
           Lake or Reservoir
- - - - - International Boundary
Boundaries are not necessarily authoritative.

0 50 100 150 200 250
Kilometers

N

This map was produced by the cartography unit of the Asian Development Bank.
The boundaries, colors, denominations, and any other information shown on this
map do not imply, on the part of the Asian Development Bank, any judgment on
the legal status of any territory, or any endorsement or acceptance of such
boundaries, colors, denominations, or information.

Source: Asian Development Bank.

The Central Asian Internal Basin, which is the largest of the three continental basins, contains rivers that supply water to the *aimags* in the water-scarce, semi-desert area. Rivers located in the western *aimags* flow into the internally drained Great Lakes Depression. River runoff is crucial for the sustenance of these lakes, of which several are ecologically important.

River discharges vary significantly from year to year and fluctuate throughout the year. Snow and ice melts are the main sources of flow during spring. The largest flow is generated by rainfall in July and August. After September, the flows decrease substantially and, in November, the water starts to freeze and the flow stops until the spring melt. Year-to-year variability is large, and important rivers such as the Tuul and the Orkhon rivers show extended periods (from 5 years to more than 10 years) of above or below long-term (i.e., more than 20 years) average flows.

Because of the variability of river flows and since rivers freeze in winter, groundwater is tapped as Mongolia's main water source (99%) for drinking and industrial water. Livestock also relies on groundwater from wells in areas away from rivers. Likewise, a majority of the country's mines uses groundwater. For irrigation, surface water is primarily used, but the use of groundwater is also increasing. A decline in groundwater levels has been observed, most likely caused by increased groundwater use. But a lack of good data on the groundwater resources and their quality makes conclusions on long-term trends and causes difficult. As these resources are economically and socially important, data collection should be improved.

Climate studies reveal that the average annual temperature has increased considerably (by about 2.2°C) since the late 1950s. In these climate studies, no long-term trend in precipitation change was detected and the observed variability of river flows seems unrelated to climate change, possibly because an observed drying climate trend is now being reversed by melting glaciers and projected increase in precipitation.[11] Still, the increase in temperature will impact the permafrost, increase evaporation and evapotranspiration, and influence the hydrology.

Trends in river flows cannot be accurately confirmed. However, a 2015 study showed indications of a reduction in lake areas on the Mongolian Plateau from the late 1980s to 2010.[12] The study suggested that Mongolia's precipitation was the primary driver of lake changes, but it also noted that the rate of loss in lake areas was greater in the PRC's Inner Mongolia Autonomous Region, where coal mining and irrigation were important factors. The study concluded that the observed decline of lake areas is likely to continue as a result of climate change and the increasing exploitation of underground mineral and groundwater resources on the Mongolian Plateau.

Although ground-based measurements of groundwater resources in Mongolia are insufficient to assess trends, the use of satellite measurement of changes in terrestrial water storage (TWS) indicates some remarkable temporal and regional variations. In the Altai Mountains, the Great Lakes Depression (also known as the Great Lakes Hollow), and the northern region of the PRC, continuous decrease in TWS can be found, which can be attributed to glacier retreat, thereby lowering the water levels of lakes and groundwater. TWS has been continuously decreasing in central Mongolia, most notably in Ulaanbaatar. In the southeast, TWS has been fairly constant.

Most river water is suitable for any use, but water pollution has become a major local issue. The rivers are heavily polluted by domestic, livestock-related, and industrial wastewater discharges. Water pollution is mainly a problem downstream of urban areas, such as in Ulaanbaatar and in *aimag* and *soum* centers. Wastewater treatment plants

[11] ADB and World Bank. 2019. Climate Change Impacts on Natural Resources—Water. In *Climate Risk Country Profile: Mongolia.* Manila. p. 14.
[12] S. Tao et al. 2015. Rapid Loss of Lakes on the Mongolian Plateau. *Proceedings of the National Academy of Sciences of the United States of America.* 112 (7). pp. 2281–2286.

are limited and often in poor condition. Mining activities also cause pollution of heavy metals, especially from small mining operations.

Surface water forms an important habitat for many bird species in the sometimes harsh (dry, cold) conditions in Mongolia. Lakes and wetlands are vital breeding grounds for seasonal birds. Hence, the environmental flow requirements in Mongolia are set high at 90%–95% of the long-term average flows. Environmental flow, which comprises the amount of water (e.g., within river basins, lakes, or wetlands) allotted to sustain ecosystem functioning and human livelihoods, is an essential element of integrated water resources management (IWRM).

2.2 Socioeconomic System

Water resources management aims to support the socioeconomic development of the country by providing water where it is needed, protecting the people from extreme climatic events (floods, droughts), and ensuring that the people are living in a healthy and enjoyable environment. Socioeconomic development specifies the conditions with which water resources management must comply. Demographic development (population growth and urbanization) and growth in the manufacturing industry need to be supported by providing the water needed for domestic and industrial use and by treating the wastewater they produce. In addition, special attention needs to be given to providing safe drinking water and sanitation facilities to the large group of herders that roam Mongolia's plains, including water for their livestock.

Water use must be distinguished between green water (water from precipitation that is stored in the root zone of the soil and evaporated, transpired, or incorporated by plants) and blue water (water sourced from surface or groundwater resources); blue water is transportable for domestic, urban, or industry use, or is consumed by the environment or agriculture). The agriculture sector in Mongolia is entirely rainfed, meaning it only uses green water.

Total water withdrawal has increased from 534 million m^3 per year in 2014 to 560 million m^3 per year in 2018. Agriculture is the biggest user of water (40%), followed by industry (25%), livestock (19%), and domestic (16%) sectors. Most of the water withdrawn by the industry and domestic sectors will be returned to the system as wastewater and (after treatment) is available for reuse. In contrast, most of the water used in agriculture is lost by evapotranspiration of plants.

The spatial variation of water withdrawals from the 29 river basins is illustrated in Figure 2. Almost half of the national domestic water supply occurs in the Tuul River because of the large population of Ulaanbaatar.

Figure 3 illustrates the total water withdrawal as a percentage of total renewable water resources by river basin. With total water withdrawal making up only 1.5% of total renewable water resources, Mongolia is reasonably well positioned to meet both current and 2030 water demands. The estimated 65% increase in water demand by 2030 to 884 million m^3 per year would require 2.5% of the total annual renewable water resources. On the local level, however, there are significant spatial differences in water demand and available water resources, resulting in local shortages. Environmental flow requirements regulate river abstraction to not exceed 5%–10% of long-term average flows to maintain ecological integrity (Section 2.1).

In most of the river basins, total water withdrawal is less than 10% of long-term average flows. However, withdrawals from the Taatsiin Tsagaan, Kharaa, and Galba–Uush–Doloodiin Govi river basins already exceed the

10% threshold—i.e., the environmental flow requirement is not maintained, which impacts ecosystem integrity and, thus, ecosystem service provision.[13] Further, this indicates that—as surface water availability is considerably reduced when maintaining environmental flows—it is likely there will be constraints in balancing the needs of socioeconomic growth and environmental sustainability in the future, especially in the low water months of dry years.

Figure 2: Water Withdrawal by River Basin and Sector, 2014
(million cubic meters per year)

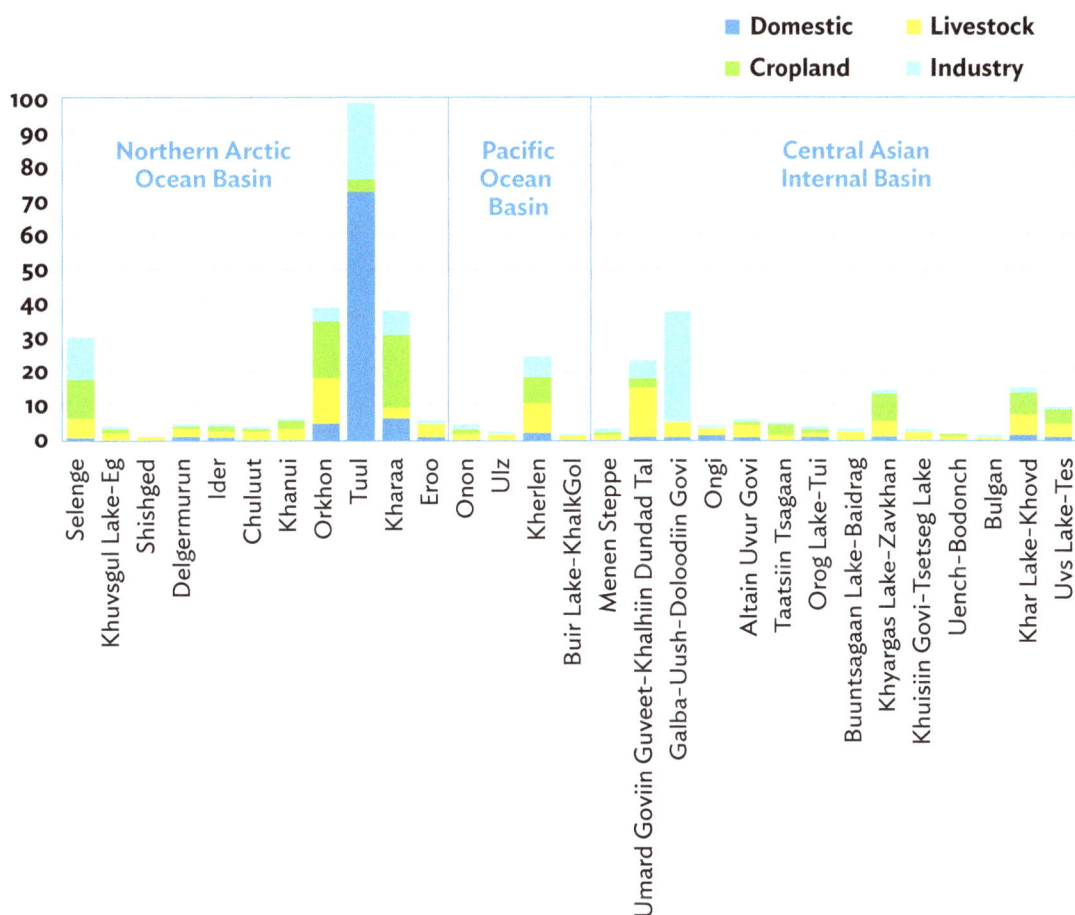

Source: Asian Development Bank. 2017. *Country Water Security Assessment of Mongolia*. Consultant's report. Manila (TA 8855-MON).

---

[13]   In Mongolia, environmental flow requirement is calculated using the percentages established for all river basins under the "surface water monograph." G. Davaa, ed. 2015. *Surface Water Regime and Resources of Mongolia*. Ulaanbaatar: Institute of Meteorology and Hydrology (in Mongolian).

Figure 3: Total Water Withdrawal as a Percentage of Total Renewable Water Resources by River Basin

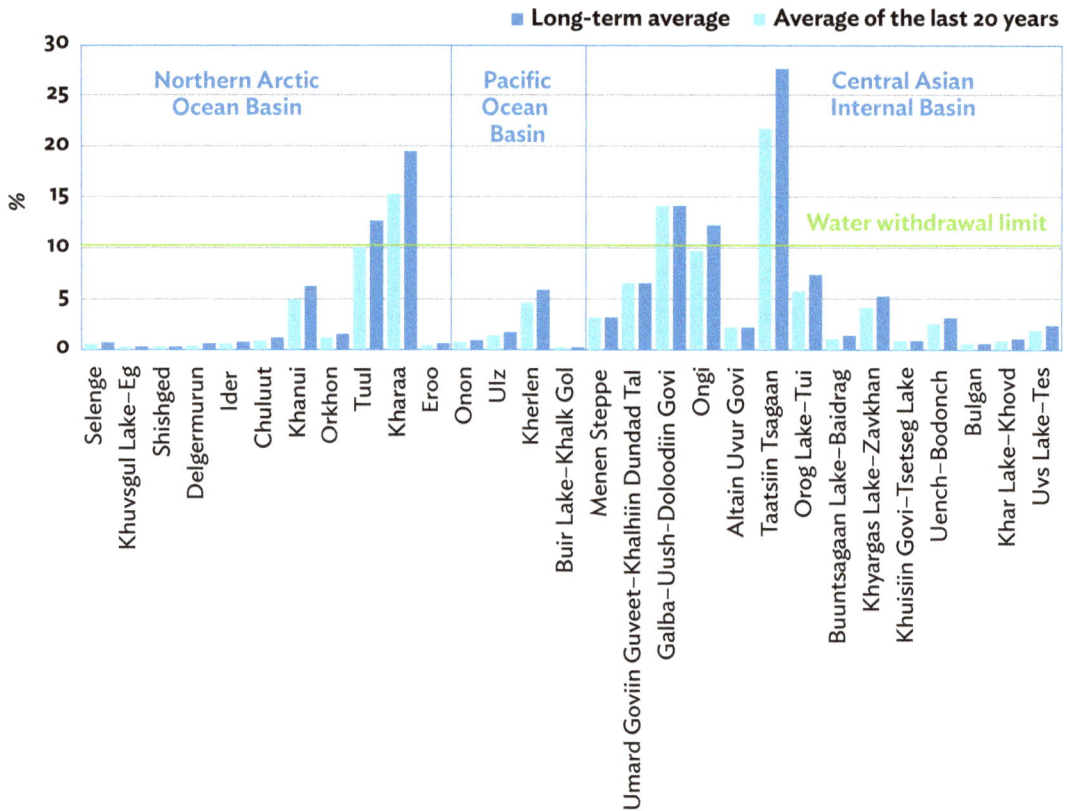

Note: Long-term average refers to more than 20 years, and average of the last 20 years refers to the period 1994–2014.
Source: Asian Development Bank. 2017. *Country Water Security Assessment of Mongolia*. Consultant's report. Manila (TA 8855-MON).

2.3 Administrative and Institutional System

For long-term sustainability and water security, the enabling conditions for good water governance need to be strong and integrated.[14] These conditions relate to national and regional laws and regulations, and they also refer to the strong institutions required to develop and implement the water resources plans, manage the system, and enforce regulations. This is a global challenge and, in Mongolia, many sectors and government departments are involved. Since water crosses administrative borders, cooperation is needed between regional administrations (e.g., *aimags* and *soums*). The preferred management for water is at river basin level.

IWRM requires the involvement of all stakeholders in the planning and management of the water resources system. This makes the institutional setting for water management complex. At the central government level in Mongolia,

---

[14] The Organisation for Economic Co-operation and Development (OECD) defines water governance as "the set of rules, practices, and processes through which decisions for the management of water resources and services are taken and implemented, and decision makers are held accountable." OECD. OECD Water Governance Programme.

five ministries are tasked with major roles in the management and exploitation of water resources. The Ministry of Environment and Tourism is the key agency responsible for integrated planning and management at the national and river basin levels, with emphasis on the environment, forestry, and water resources. The other ministries involved and their responsibilities are as follows: (i) the Ministry of Energy for activities such as cooling water power stations and hydropower; (ii) the Ministry of Health for water hygiene and water-related diseases; (iii) the Ministry of Food, Agriculture and Light Industry for agriculture (including irrigation), livestock, and industry; and (iv) the Ministry of Construction and Urban Development for urban water supply, flood protection, and major construction projects. Coordination at the central government level is handled by the National Water Committee (cross-sector coordination), the Water Resource Council (water resources research, exploration, and estimation results approval and review), and the Water Services Regulatory Commission (urban water tariff and licensing for providers).

Mongolia's main policy documents on water include the National Water Program (2010), National Security Concept (2010), IWMP (2013), Green Development Policy (2013), and Mongolia Sustainable Development Vision (SDV) 2030 (2016). The IWMP 2013 is based on the principle of IWRM, specifies the overall objectives of water management in the country in terms of water security, and outlines the strategy to achieve water security.

Decentralized Decision-Making

Under the ratified Constitution of 1992, lower administrative units can organize their own functions and responsibilities. This facilitates the application of direct solutions to local-scale issues, which are locally resolved under the supervision of the central government. The Integrated Budget Law, launched in 2013, supports the decentralization process by granting local governments with discretionary powers to use local development funds. Decentralization paves the path for local commitment on development initiatives and strengthens the role of water and environmental management at the *soum* level.

The 2004 Water Law initiated the reassignment of many water functions from the central government to local governments, with *aimag* and *soum* governors responsible for matters related to the use, protection, and restoration of water resources. River basin councils (RBCs) have been established since 2009 and act as forums for water resources planning and management for all the stakeholders and districts within a river basin. The 2012 Water Law introduced several new concepts important for IWRM, including the following:

(i)   defining the mandates of state organizations responsible for the development and adoption of IWRM plans;

(ii)   introducing river basin organizations (RBOs) in Mongolia's 29 river basins that work in parallel with each other, thereby paving the way for the decentralization of water management and facilitating the involvement of citizens in water management;

(iii)   providing the legal basis for the introduction of IWRM and establishing an institutional framework, including a firmer positioning of essential IWRM organizations; and

(iv)   opening the way for the active engagement of the private sector in water management activities, including state corporations and public–private partnership undertakings.

Based on the 2012 Water Law, 29 RBOs will eventually be established and each RBO comprises two components: (i) a river basin authority (RBA) composed of 5–12 government personnel reporting directly to the River Basin Management Division of the Ministry of Environment and Tourism; and (ii) an RBC consisting of about 35 members with representatives from government, stakeholders, and water users. The key tasks of the RBOs are to develop and implement river basin management plans through multisector stakeholder consultations.

As new elements in Mongolia's government structure, RBOs will need time to mature and function optimally. Given limitations in resources and capacity, only 21 RBOs have so far been established. Figure 4 gives an overview of the institutional setting of IWRM in Mongolia, including the RBO's position.

Figure 4: Overview of the Administrative and Institutional System for Integrated Water Resources Management in Mongolia

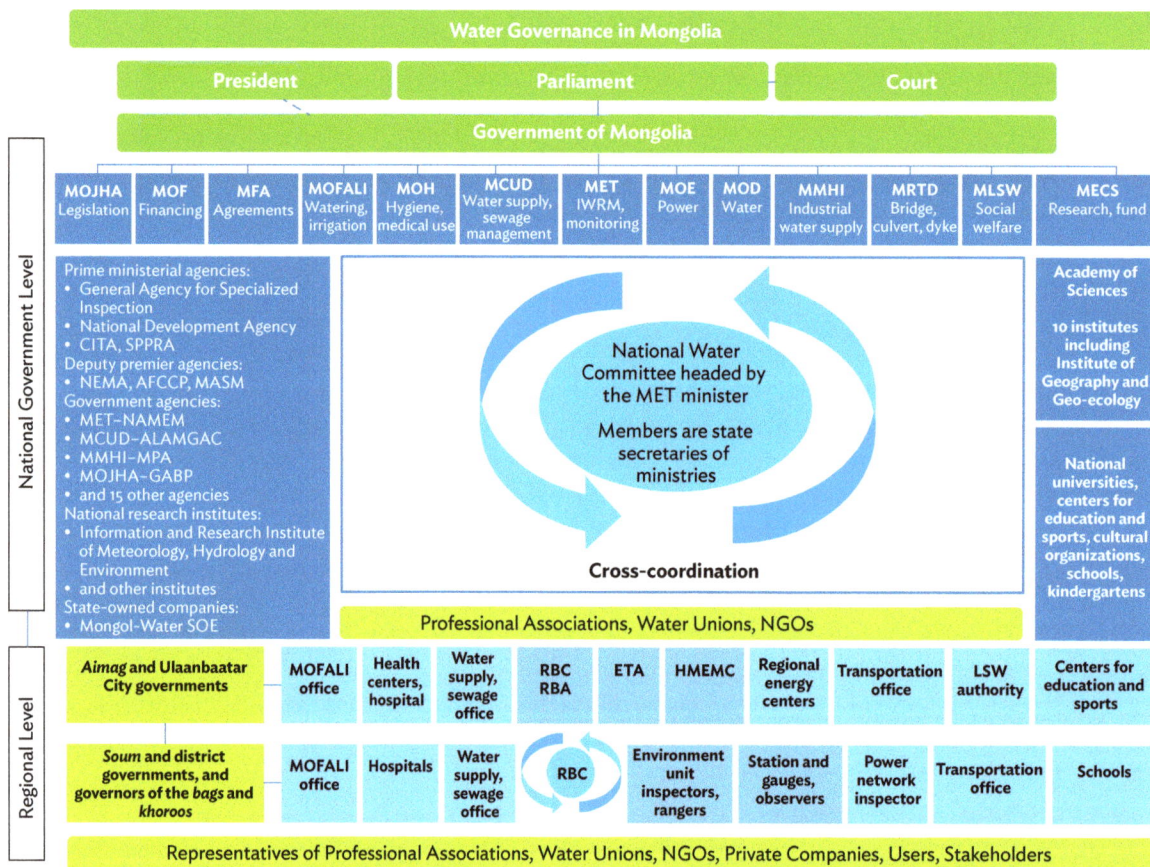

AFCCP = Agency for Fair Competition and Consumer Protection; ALAMGAC = Agency for Land Administration and Management, Geodesy and Cartography; CITA = Communication and Information Technology Agency; ETA = Environment and Tourism Authority; GABP = General Agency for Border Protection; HMEMC = Hydrology, Meteorology and Environment Monitoring Center; IWRM = integrated water resources management; LSW = labor and social welfare; MASM = Mongolian Agency for Standardization and Metrology; MCUD = Ministry of Construction and Urban Development; MECS = Ministry of Education, Culture and Science; MET = Ministry of Environment and Tourism; MFA = Ministry of Foreign Affairs; MLSW = Ministry of Labor and Social Welfare; MMHI = Ministry of Mining and Heavy Industry; MOD = Ministry of Defense; MOE = Ministry of Energy; MOF = Ministry of Finance; MOFALI = Ministry of Food, Agriculture and Light Industry; MOH = Ministry of Health; MOJHA = Ministry of Justice and Home Affairs; MPA = Mineral and Petroleum Agency; MRTD = Ministry of Road and Transport Development; NAMEM = National Agency for Meteorology and Environment Monitoring; NEMA = National Emergency Management Agency; NGO = nongovernment organization; RBA = river basin authority; RBC = river basin council; SOE = state-owned enterprise; SPPRA = State Property Policy and Regulation Agency.

Notes:

1. MET is the key agency responsible for IWRM, planning, and monitoring, with emphasis on the environment, forestry, and water resources.
2. The administrative divisions of Mongolia are *aimag* (province), *soum* (district), *bag* (subdistrict), and *khoroo* (municipal subdistrict).

Source: Asian Development Bank.

3 Water Management in Mongolia: History, Challenges, Trends, and Future

Based on long-term average actual renewable water resources, Mongolia has enough water, by volume, to support its population and economic activities, both at present and in the future. However, the main issue is that this water is not available at the right place, at the right moment, and in the right quality. Infrastructure needed to ensure water supply security to water consumers is far from adequate. Although the institutional structures to develop and manage water resources are being established, they have yet to fully adapt to the new market economy (i.e., after the socialist centralized system was abandoned in the early 1990s). Government administration for water nearly collapsed when that shift was made. Water is a common good, and free market principles do not apply to common goods. Without proper government regulations, most consumers feel free to use water as they wish and free-rider situations will appear. The government is, therefore, developing institutions, legislation, and regulations to provide the right incentives for consumers to use water in a responsible way in a market-oriented society.

3.1 Implementing Integrated Water Resources Management

The Government of Mongolia is rebuilding the way water is managed in the country, following the integrated water resources management (IWRM) approach as the leading principle. IWRM is specified in the 2012 Water Law of Mongolia and was also the basic approach used in developing the national Integrated Water Management Plan (IWMP) of 2013. The IWRM concept has been developing since the 1980s in response to the mounting pressure on water resources systems all over the world, and it is now globally accepted as the way forward in water management. Shortages in water supply and decline in water quality have compelled many countries to reexamine their development policies on water resources management. Consequently, the management of water resources has seen a global transformation from a largely supply-oriented, engineering-biased approach to a demand-oriented, multisector approach, now labeled as IWRM.[15] IWRM addresses not only the natural resources system, but also the socioeconomic system and the administrative and institutional system (Figure 5).

The IWRM concept shifts from a "top-down water master planning" that focuses on the development and availability of water resources toward a "comprehensive water policy planning" that facilitates multisector interaction, sets priorities, takes into account institutional requirements, and promotes the enhancement of management capacity (footnote 15). IWRM links water resources consumption to social and economic purposes

---

[15] D. P. Loucks and E. van Beek. 2005. *Water Resources Systems Planning and Management: An Introduction to Methods, Models and Applications.* Paris: United Nations Educational, Scientific and Cultural Organization.

and pursuits. As such, it influences the regulations and legislation, as well as the infrastructure, needed to promote the sustainable and effective use of water resources, including meeting ecosystem needs. IWRM is a process in which all stakeholders jointly decide on how to develop and manage their water resources system.[16]

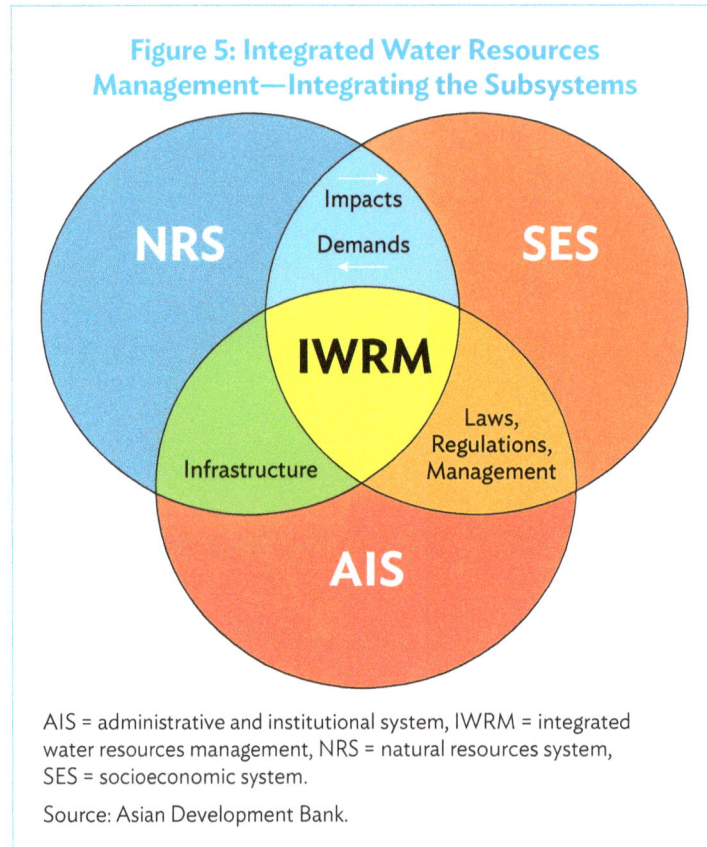

Figure 5: Integrated Water Resources Management—Integrating the Subsystems

AIS = administrative and institutional system, IWRM = integrated water resources management, NRS = natural resources system, SES = socioeconomic system.

Source: Asian Development Bank.

Crucial to the implementation of IWRM are its enabling conditions or the "three pillars of IWRM" (Figure 6):

(i)   **Enabling environment.** Water-related national policies, regulations, and legislation that guide planning and enable enforcement.

(ii)   **Institutional framework.** Existence of water institutions with qualified staff at the national and regional levels, and some form of a river basin organization (RBO) at the river basin level.

(iii) **Management instruments.** Availability of and access to information, data, and tools that enable informed decision-making.

An important feature of IWRM is sustainability. The vision of Mongolia on sustainable development is stated in the parliamentary document Mongolia Sustainable Development Vision (SDV) 2030.[17] IWRM is mentioned as

---

[16]   IWRM is a process that fosters the coordinated development and management of water, land, and associated resources to equitably maximize the ensuing social and economic benefits without compromising the sustainability of essential ecosystems. Global Water Partnership Central and Eastern Europe. What is IWRM?

[17]   Parliament of Mongolia. 2016. *Mongolia Sustainable Development Vision 2030*. Ulaanbaatar.

one of the three components of the SDV 2030's strategic objective 3 on environmental sustainability, with two specific goals: (i) protect water resources and prevent water shortage; and (ii) increase drinking water supply that meets established quality standards, and increase accessibility to improved hygiene and sanitation facilities. These goals are quantified in terms of clear targets for 2020, 2025, and 2030. The other two components of environmental sustainability are (i) coping with climate change and (ii) ecosystem balance. Both are also very important for water management.

Figure 6: The Three Pillars of Integrated Water Resources Management

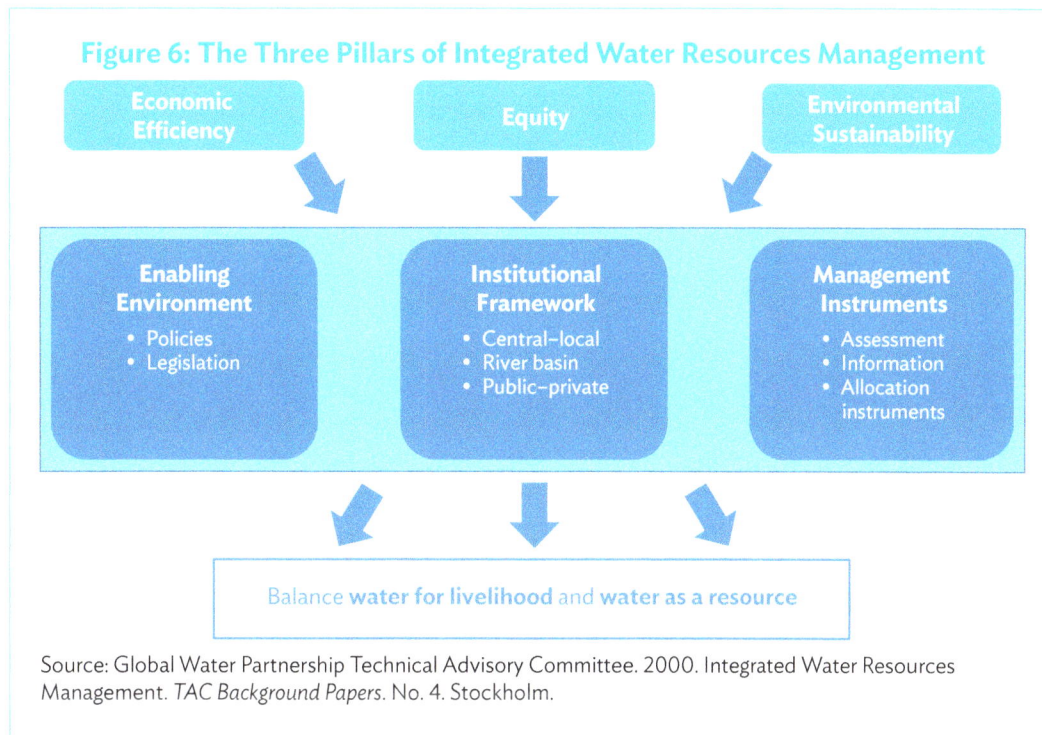

Source: Global Water Partnership Technical Advisory Committee. 2000. Integrated Water Resources Management. *TAC Background Papers*. No. 4. Stockholm.

The Sustainable Development Goals (SDGs), as adopted by the United Nations in 2015 with the 2030 Agenda for Sustainable Development, have strong links with the strategic objectives of the SDV 2030, although there are differences in targets between the general SDGs and the SDV 2030. For example, whereas the SDGs strive for universal access to safe drinking water for all, Mongolia's SDV 2030 strives to ensure that 90% of the population is supplied with safe drinking water by 2030. The differences reflect the special conditions in Mongolia and the magnitude of the challenges that Mongolia faces.

Basin Health Report Card

The Tuul RBA, together with the World Wide Fund for Nature (WWF) and the University of Maryland Center for Environmental Science, developed a methodology for a river basin health report card for Mongolia and applied this to the Tuul River Basin (Box 1). Although much research has been conducted on the water quality and quantity of the Tuul River, there is no standardized or consistent assessment on the state of the basin. Thus, the report card provides the first consolidated baseline assessment of basin health. It includes economic, environmental, and social values that can be tracked.

Box 1: Tuul River Basin Health Report Card

Report cards are assessment and communication products that compare economic, ecological, and/or social information against predefined targets. They effectively synthesize large and often complex information into simple scores that can be shared with decision makers and the general public. The process of developing report cards is highly participatory and includes the (i) identification of values and threats, (ii) selection of indicators, (iii) definition of thresholds, (iv) calculation of scores, and (v) communication of results.

The Tuul River Basin Health Report Card was created through a series of stakeholder workshops with representatives from Mongolian government agencies, academic institutions, nongovernment organizations, and the private sector. The Tuul River Basin was divided into six regions based on ecosystem condition, urbanization, socioeconomic development, water use, and water pollution. The assessment was then conducted for each indicator at the regional and river basin levels. The first report card serves as a baseline to measure changes in the future in response to management actions, inform policy and planning within the basin, and assist in revising the 2012 Tuul River Basin Integrated Water Management Plan.[a] The figure shows the basin health indicators identified for the Tuul River.

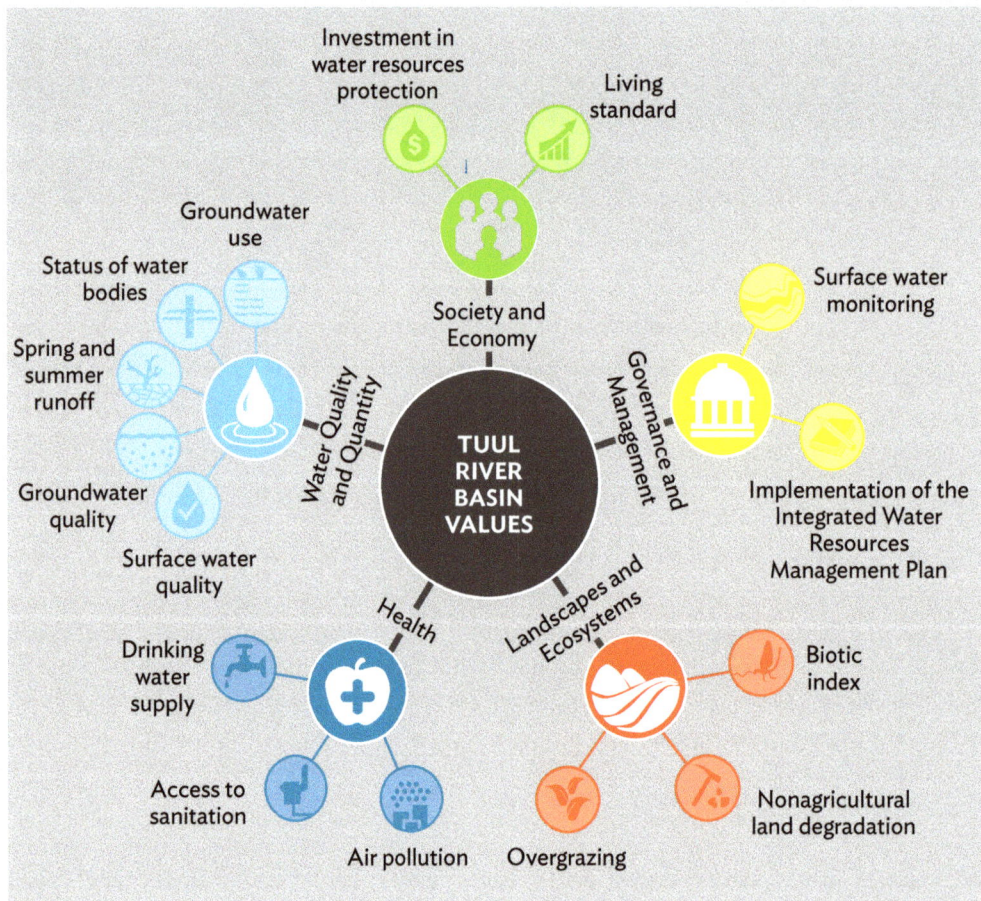

continued on next page

Box 1 *continued*

The grading system of the basin health report card is defined as follows:

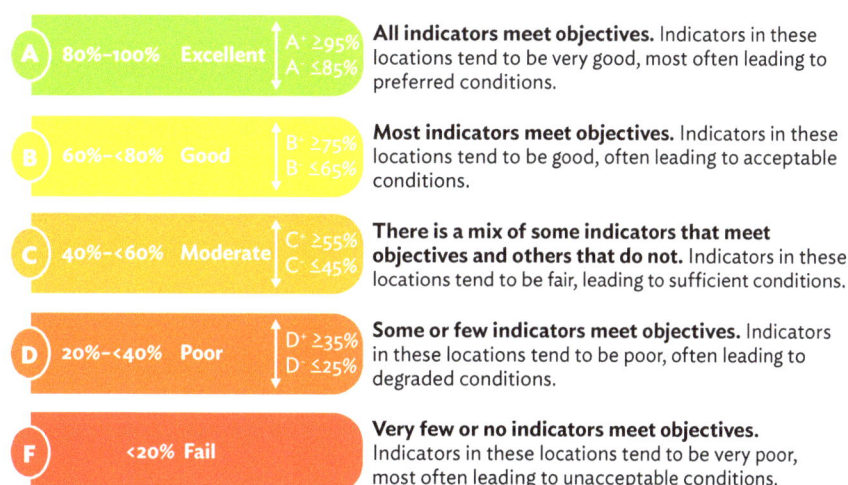

A	80%–100% Excellent	A⁺ ≥95% / A⁻ ≤85%	**All indicators meet objectives.** Indicators in these locations tend to be very good, most often leading to preferred conditions.
B	60%–<80% Good	B⁺ ≥75% / B⁻ ≤65%	**Most indicators meet objectives.** Indicators in these locations tend to be good, often leading to acceptable conditions.
C	40%–<60% Moderate	C⁺ ≥55% / C⁻ ≤45%	**There is a mix of some indicators that meet objectives and others that do not.** Indicators in these locations tend to be fair, leading to sufficient conditions.
D	20%–<40% Poor	D⁺ ≥35% / D⁻ ≤25%	**Some or few indicators meet objectives.** Indicators in these locations tend to be poor, often leading to degraded conditions.
F	<20% Fail		**Very few or no indicators meet objectives.** Indicators in these locations tend to be very poor, most often leading to unacceptable conditions.

Overall, the Tuul River Basin scored 49% (i.e., *C* or *moderate*). Indicators varied widely in their scores with "air pollution" achieving an *F* (*fail*), and "access to drinking water supply" and "spring and summer runoff" each achieving an *A* (*excellent*). In general, indicators under the "water quality and quantity" category scored best, while those under the "governance and management" category scored worst. The report card makes a series of recommendations to improve the health of the Tuul River.

ᵃ Ministry of Environment and Green Development. 2012. *Tuul River Basin Integrated Water Management Plan*. Ulaanbaatar.
Source: WWF Mongolia Office; Government of Mongolia, Ministry of Environment and Tourism; Tuul River Basin Authority; University of Maryland Center for Environmental Science; and Asian Development Bank. 2019. *Tuul River Basin Health Report Card 2019*. Washington, DC.

3.2 Water Challenges and Hot Spots

Mongolia's economy has recovered sharply from the 2015–2016 downturn. Its gross domestic product increased to 5.3% in 2017 and 7.2% in 2018; moreover, because of the large investments in mining, gross domestic product continued to grow, although at a slower pace, at 5.1% in 2019.[18] However, Mongolia's economy, as with the rest of the world, is expected to suffer a heavy toll in 2020 from the massive disruption brought about by the coronavirus disease (COVID-19) global pandemic. Poverty incidence remains high, with 28.4% of the population living below the national poverty line in 2018[19]—even reaching 40% during 2014–2016—along with widening inequalities, rising unemployment, and escalating environmental concerns such as air and water pollution. Mongolia also faces several water management challenges, the most urgent of which are related to drinking water supply, wastewater

---

[18] ADB. Economic Indicators for Mongolia (accessed 7 May 2020).
[19] ADB. Poverty Data for Mongolia (accessed 7 May 2020).

treatment, and bulk water provision for urban centers and the mining sector. In the longer term, climate change might have severe impacts on the country's water resources system. All these issues could undermine Mongolia's ability to achieve the SDGs.

3.2.1 Public Water Supply, Sanitation, and Wastewater Treatment

The government has made significant investments in improving the country's water management situation. In cooperation with international partners, water supply and sanitation (WSS) infrastructures have been developed through development assistance and donor-funded projects. However, the Government of Mongolia and the United Nations Economic and Social Commission for Asia and the Pacific reported the following:[20]

(i)   About 13% of the urban population and 56% of the rural population have no access to safe drinking water. Insufficient access to safe drinking water has a direct negative impact on the health of the population.

(ii)   Wastewater treatment facilities in Ulaanbaatar and in most provincial centers are old and use outdated technologies, thereby failing to adequately treat wastewater to the required standard. When untreated or poorly treated wastewater is discharged to naturally occurring water sources, such as the Tuul River, it causes pollution and brings health hazards to humans and animals.

The Government of Mongolia is struggling to address the upsurge in the demand for water and sanitation facilities, as well as the increase in the volume of wastewater—both the result of rapid urbanization. There is no system in place for the reuse or recycling of treated wastewater. Although the government promulgated policies to promote the reuse of treated wastewater, the water pricing system discourages the introduction or application of recycling technologies and practices. This is because the cost of reusing treated water is higher than using clean water. Likewise, with the amount of wastewater exceeding the capacity of water treatment facilities, surface water contamination has increased. Without appropriate actions, this situation is expected to further deteriorate. The main hot spot for WSS and wastewater treatment is Ulaanbaatar. But, on a smaller scale, the same issues need to be addressed in other urban centers at the *aimag* and *soum* levels.

3.2.2 Bulk Water Supply and the Mining Sector

Although Mongolia is well endowed with water (both per capita and nationally), the concentration of the population in a few urban centers and the development of economic activities (mining) in the dry southern regions have made water availability an issue. Potential solutions might include increasing bulk water supply by means of storage and interbasin transfers, although these may have negative effects on the source basins, including international implications. Where possible, solutions should be found in reducing water demand— e.g., through water reuse or through application of less water-intensive production processes. The main hot spots for bulk water supply are Ulaanbaatar and the mining sector in the Gobi Desert region. Both hot spots have been addressed in detailed studies by the 2030 Water Resources Group (footnote 6).

---

[20]   United Nations Economic and Social Commission for Asia and the Pacific; and Government of Mongolia, Ministry of Environment and Tourism, National Development Agency, and National Statistics Office. 2018. *Sustainability Outlook of Mongolia*. Ulaanbaatar.

3.2.3 Climate Change-Related Issues

Climate change is expected to become a major challenge for Mongolia. Based on measurements from 48 meteorological stations distributed across the country, Mongolia's temperature increased by 2.2°C from 1940 to 2015. Assessment of these data reveals higher intensity warming in the mountainous regions and lower intensity in the steppe and *gobi* regions. Short-term projections for 2016–2035 show seasonal temperature increases of 2.0°C–2.3°C, while long-term projections for 2081–2100 range from 2.4°C to 6.3°C, depending on the representative concentration pathway scenarios used (footnote 10). Because of its geographical and climatic conditions, Mongolia is considered as one of the most vulnerable countries to the impacts of climate change. Such vulnerability is further aggravated by the level of development and structure of its economic sectors, and the living standard of its population. The rise in temperature will have direct impacts on the country's soil, pasture, forest ecosystems, and fauna. The resulting land degradation and desertification will impact animal husbandry and arable farming. With respect to the impacts on water management, the conclusions are somewhat less pronounced and include the following:

(i)   Since the 1940s, there has been a very small decrease in annual precipitation and a small increase in winter precipitation. Most of the climate models indicate that Mongolia's precipitation will slightly increase by 20 millimeters by 2050 and 44 millimeters by 2090, with the majority of the increase predicted to happen during the winter months.

(ii)   The anticipated annual increase in precipitation will be accompanied by greater seasonal variability leading to floods and droughts.

(iii)   An estimated 28% loss of glacier volume has been recorded since 1940. Although glacier melt contributes to increasing the quantity of river water, the steady decline of glaciers will intensify the variability of available water. Most studies indicate that glacier melt will peak during 2030–2050 and decline thereafter.

(iv)   Projections indicate that annual river flow will decrease by 2% for every 1°C increase in temperature.

A reduction of river flows was observed during 1996–2016. However, it cannot be ascertained whether such reduction signifies a long-term trend or is just a manifestation of the cyclical nature of climate. Mongolia's 2013 national IWMP assessed that river flow reduction appeared to be within the realm of a normal variability in climate rather than a result of climate change.

Water resources management should be strengthened in anticipation of possible stresses to water users and the water environment arising from climate change. What is needed initially is to continue with the extensive monitoring and analysis of the changes and to carry out studies on the potential impacts on the water resources system, including glaciers and permafrost. In parallel, programs for adaptation to projected climate change impacts should be further developed.

Emphasis could be given to adaptation in the livestock industry, where many herders have limited resources to respond to production shocks. Mongolia's Third National Communication under the United Nations Framework Convention on Climate Change (footnote 10) states that climate change projections have shown expected increases in the intensity of summer drought and harsh winter or *dzud* (Box 2) in the country. Livestock loss is projected to increase by about 50% by 2050 and 100% by 2100. In addition, climate change could exacerbate livestock diseases and is projected to impact on the security of water sources used by herders as the "area and volume of glaciers and number of small lakes and ponds in high-mountains are much reduced" and an "increasing trend of evaporation predicts entirely dry conditions".[21] Other adaptation needs during 2021–2030 are briefly described in Mongolia's Intended Nationally Determined Contribution.[22]

Box 2: What is a *Dzud*?

A *dzud* is a complex meteorological natural phenomenon, where a severe winter (November–February)—characterized by abnormally low temperatures and high winds—occurs after a dry summer. Summer drought (July–September) often results in withered grasslands and, therefore, insufficient grazing for livestock. With a limited food source, livestock become underweight and less resilient to endure the coming harsh winter. Hence, *dzuds* cause overwhelming loss of livestock, leading to economic crisis and food security issues. There are various forms of *dzuds*, as described below.

Type of *Dzud*	Condition	Effect
Tsagaan (white)	This is the most common form of *dzud*, where a thick layer of snow covers the pasture land.	Affects large areas and prevents animals from grazing
Har (black)	Winter has no snow, yet temperatures are consistently colder than average.	Causes water shortage and usually coincides with scarcity of winter grass
Tumur (iron)	Snowmelt freezes, covering the pasture land with ice.	Obstructs access to pasture, hindering livestock from grazing
Tuurain (hoof)	The large number of livestock (i.e., the abundance of hoofs) has exceeded the carrying capacity of the pasture land resulting in overgrazing and land degradation.	Depletes grassland and limits grazing area for livestock
Havsarsan (combined)	This is a combination of any of the above conditions.	Compounds above effects, exacerbating the strain on livestock

Source: P. Batimaa and L. Natsagdorj. 2008. Drought, *Dzud* and Climate Change in Mongolia. In N. Leary et al, eds. *Climate Change and Vulnerability*. London: Earthscan. pp. 196–210.

3.2.4 Transboundary Water Management

Few small rivers flow into Mongolia, while about 210 bigger rivers flow out of the country into the People's Republic of China (PRC) and the Russian Federation. The area draining into the Russian Federation encompasses about one-third of Mongolia's territory. The Government of Mongolia has entered into transboundary agreements for the protection, utilization, and pollution control of transboundary waters with the governments of the PRC and the Russian Federation.

The agreement with the Government of the PRC incorporates areas of cooperation, which cover the following:

(i)   investigation and survey of dynamics, resources, and quality of boundary waters;

(ii)   monitoring of transboundary waters and protection of aquatic animals; and

(iii) development of approaches for the breeding and protection of fish resources.

---

[21] Footnote 10, pp. 199–200.
[22] Footnote 9, Table 6.1, pp. 41–42.

The broad areas of cooperation under the agreement with the Government of the Russian Federation include the following:

(i)   rational use of transboundary waters, including environmental flows;

(ii)   prevention of pollution and monitoring of water quality;

(iii)   examination of transboundary water resources, including biology and water chemistry;

(iv)   protection of fish and birds, including the natural environment for migratory birds;

(v)   exchange of information, including on flood and industrial calamities; and

(vi)   continued cooperation, including integration of methodology, for defining and monitoring environmental flows, and for developing standards and rules for transboundary water allocation and utilization.

4 Water Security Assessment of Mongolia

Mongolia's country water security assessment (CWSA), which was conducted at the national and river basin levels (footnote 4), provides a quantitative and comprehensive review of the water security condition in Mongolia, with the goal of initiating dialogue and targeted action to move toward a water-secure future. The CWSA is based on the *Asian Water Development Outlook* (AWDO) 2016 methodology,[23] which was developed by ADB and the Asia-Pacific Water Forum. This methodology was customized to account for the Mongolian context, allowing for more targeted analysis. The adjustments made to the AWDO methodology were based on stakeholder consultations and the results of applying the methodology to five pilot river basins.

4.1 Methodology

ADB introduced and defined the concept of a water security index, during the first Asia-Pacific Water Summit held in Japan on 3–4 December 2007, as part of the preparation for the AWDO 2007 publication.[24] In 2013, the AWDO approach was further developed to include quantitative measurements and provide guidance for governance, capacity building, investments, monitoring, and reporting.[25] It was further revised in 2016, with the AWDO 2016 assessing the water security of 48 Asian countries, including Mongolia (footnote 23).

A water security framework was established under the AWDO 2016 based on five key dimensions (KDs): household water security (KD1), economic water security (KD2), urban water security (KD3), environmental water security (KD4), and resilience to water-related disasters (KD5). Each KD is quantified by measurable indicators (Figure 1).

While each KD allows for insight into the water security of its respective dimension, the five KDs are interdependent. Assessments of overall water security and decisions in the water sector can only be rendered effectively when all KDs are combined.

Water security—be it at the national or river basin level—is, therefore, assessed as a composite result of these five KDs. Each KD is rated on a scale from 1 (*high risk* or *low water security*) to 20 (*low risk* or *high water security*). The score for the overall national or river basin water security is the sum of the ratings of the five KDs, with a maximum of 100 points. In the case of water security assessment at the river basin level, either a population-weighted or an unweighted average of the river basin-level scores are taken to derive the national water security

[23] ADB. 2016. *Asian Water Development Outlook 2016: Strengthening Water Security in Asia and the Pacific.* Manila.
[24] ADB. 2007. *Asian Water Development Outlook 2007: Achieving Water Security for Asia.* Manila.
[25] ADB. 2013. *Asian Water Development Outlook 2013: Measuring Water Security in Asia and the Pacific.* Manila.

index. The total score determines the water security stage, with the five stages ranging from *hazardous* (worst) to *model* (best) (Table 1).

Table 1: Water Security Stages

Index	Score	Stage	Description
5	96 and above	Model	Access to safe drinking water and sanitation facilities achieved for all; water availability issues no longer constrain economic activities; water quality meets established standards for the population and the ecology; and water-related risks are tolerable and relatively well managed.
4	76 to < 96	Effective	Access to safe drinking water and sanitation facilities achieved for nearly all people; water service delivery is typically well structured and more effectively supports economic activities; water quality is generally acceptable, with adequate consideration of ecological restoration of water bodies; and water-related risks are considerably managed by both structural and nonstructural (early warning systems) measures.
3	56 to < 76	Capable	Access to safe drinking water and sanitation facilities is continually improving, including in poor and rural areas; water productivity has increased to support economic activities; water quality is getting better because of regulatory and wastewater treatment measures; initial steps for restoring the ecological health of water bodies are taken; and the most critical water-related risks are being dealt with.
2	36 to < 56	Engaged	Access to modest drinking water and sanitation facilities provided to more than half the people; water service delivery is being developed and starting to support economic activities; initial measures to improve water quality are commenced; and first attempts to address water-related risks are undertaken.
1	< 36	Hazardous	Drinking water and sanitation facilities are limited and pose serious health hazards; the mostly informal water service delivery is constraining economic activities; water quality is poor and unsafe for people; the aquatic ecology suffers serious damage; and calamities like floods and droughts push more people into poverty.

Source: Asian Development Bank. 2017. *Country Water Security Assessment of Mongolia.* Consultant's report. Manila (TA 8855-MON).

Adjusted Methodology for the Mongolian Context

Water security has been quantified in the AWDO 2016 to facilitate comparison of water security across ADB members. While the adopted methodology should ideally be applicable to all economies covered by the AWDO 2016, international and domestic variations (e.g., in level of economic development, natural resources endowments, physical conditions, social issues, and challenges) require adjustments to take into account specific local conditions while applying the AWDO methodology. For Mongolia, the AWDO's overall approach of using the five KDs remained the same, but adjustments were made on how these KDs are determined. In addition, the CWSA was applied at the river basin level to account for regional differences within Mongolia.

The CWSA was conducted in collaboration with stakeholders at the national and river basin levels. Identification of the needed adjustments to the AWDO methodology, based on the particular conditions of Mongolia (Table 2), followed these five steps:

(i) The AWDO 2016 indicators used for the estimation of the five KDs were reviewed, and the applicability of these indicators was evaluated in relation to the availability of data in Mongolia at the river basin level. From this preliminary review, some indicators were amended and new ones were added to reflect the unique circumstances in Mongolia.

(ii)   For each of the five KDs, a river basin was chosen as a pilot area to further examine the applicability of the adjusted methodology. The basis for selection is the presence of typical characteristics in the basin for the corresponding KD.

(iii) Primary and secondary data were collected and tested in the pilot river basins to fine-tune the respective indicators of the five KDs. Primary data were gathered from individual interviews, focus group meetings, surveys, and workshops at the *aimag* and *soum* levels. About 500 participants attended the meetings, while over 1,000 respondents completed the survey questionnaires. Collected data were then assessed for reliability, analyzed, and triangulated. Mongolia's adjusted methodology was finalized based on the study of the five pilot river basins.

(iv)   The adjusted methodology was applied to Mongolia's 24 remaining river basins.

(v)   Given the uneven distribution of Mongolia's population across the country, the national CWSA was calculated using the weighted average of all river basin-specific KD scores based on (a) population (rural population for KD1, urban population for KD3, and total population for KD2 and KD5) and (b) land area (for KD4). This weighted process was also applied in computing the averages of the three drainage basins (Northern Arctic Ocean Basin, Pacific Ocean Basin, and Central Asian Internal Basin).

Based on the lessons from the pilot studies, water security scores for all 29 river basins have been determined and, based on these river basin scores, an overall score for the country as well as scores for each drainage basin have been calculated (sections 4.2 and 4.3). The water security assessment was carried out with the support of staff from the river basin organizations (RBOs) who were trained in applying the CWSA methodology. Integrating capacity building into this participatory approach allows the periodic conduct of water security assessment at the river basin level and, in the future, at the national level.

Table 2: Adjustments to the Asian Water Development Outlook 2016 Methodology for Mongolia's Country Water Security Assessment

KD	Adjustment to the AWDO 2016 Methodology
Household water security (KD1)	• Focus is exclusively on rural household water security. • The indicator "improved water supply" replaces "piped water supply," which is not practicable for herder households. • An indicator on distance of herder households to main water source is added. • Scoring for sanitation is relaxed.
Economic water security (KD2)	• An indicator for livestock is added, as livestock raising is the prevalent activity in the agriculture sector and it has a different water requirement than crops.
Urban water security (KD3)	• Similar to KD1, indicators on water supply and sanitation are adjusted.
Environmental water security (KD4)	• Scoring methodology is simplified.
Resilience to water-related disasters (KD5)	• Given that Mongolia is landlocked and faces unique challenges from *dzud* (a severe winter after a summer drought), an indicator on "resilience against *dzud*" replaces "storm surges and coastal floods."

AWDO = Asian Water Development Outlook, KD = key dimension.
Source: Asian Development Bank. 2017. *Country Water Security Assessment of Mongolia.* Consultant's report. Manila (TA 8855-MON).

Consistent with the government's program for achieving the 2030 Agenda for Sustainable Development, 2014 was used as the base year for the water security assessment; therefore, the analysis used the 2014 socioeconomic data of 330 *soums*. It also used the averages for 20–30 years of climate and water resources data derived from the hydrometeorological monitoring network of Mongolia. Although water security indicators for the country and by river basin were analyzed using 2014 data, the assessment was for the most part done in 2016, adopting the amended 2016 AWDO methodology.

4.2 Water Security at the National Level

Results of the CWSA at the national level are given in Figure 7. Figure 7(a) shows the overall score for the five KDs, while figures 7(b)–7(f) present the scores by KD for the indicators identified for that KD. Figure 7(a) shows that all KDs are at a similar level, i.e., from 10 to 15. Among the five key water security dimensions, the weakest is KD1 (rural household) and the strongest is KD4 (environmental). Interestingly, average scores for KD2, KD3, and KD5 are strongly influenced by scores of the Tuul River Basin, as about half the population resides there.

The rural household water security (KD1) score is relatively low, primarily because of the low access of herder communities to improved water supply and sanitation (WSS). Also of concern is the low economic water security (KD2) score, indicating that water constraints—particularly during droughts and periods of low discharges—adversely affect economic development. This is especially the case in Mongolia's southern region, where there is full dependency on the limited supply of groundwater, the sustainability of which is jeopardized by low recharge and projected further climate change-induced reduction. The quality of some groundwater is already deteriorating, and lower recharge rates will further impact on water quality. Owing to the limited proportion of the urban population with access to piped water supply and sewerage, the score of urban water security (KD3) is likewise low. The low average scores of KD1 and KD3 render both rural and urban areas as hot spots. This means targeted investments and focused management are essential to provide improved access to affordable and reliable WSS facilities. In contrast, the score of environmental water security (KD4) is high mainly because a reasonable proportion of land has been designated as state and local protection areas, population in these areas is low, and water quality is relatively good. In addition, Mongolia's agriculture sector has low fertilizer usage, which is beneficial to the environment. Similarly, the score of resilience to water-related disasters (KD5) is satisfactory. Table 3 presents the five KDs of water security, with their corresponding scores and core interventions.

Figure 7: Mongolia's National Water Security Scores by Indicators per Key Dimension

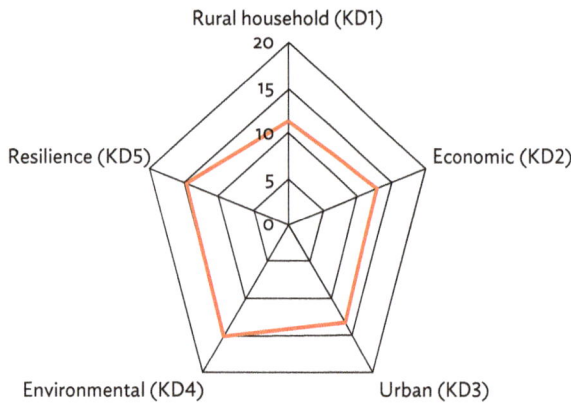

(a) Overall National Water Security

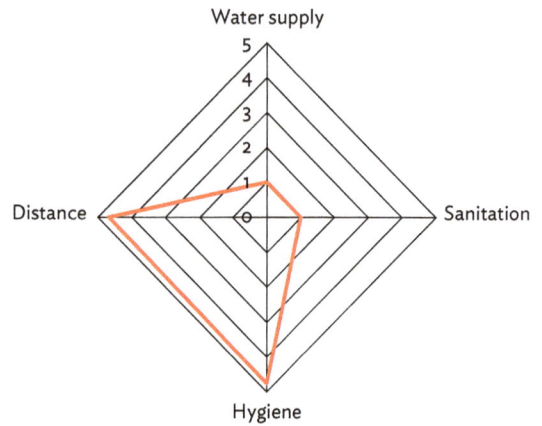

(b) Rural Household Water Security (KD1)

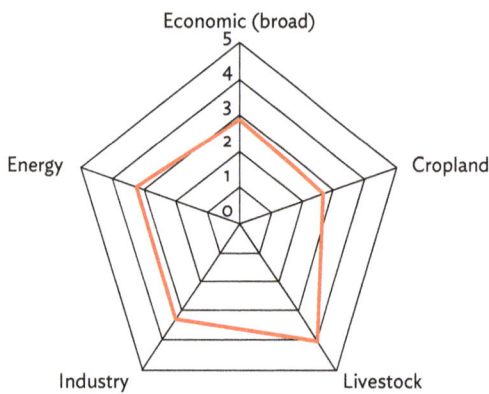

(c) Economic Water Security (KD2)

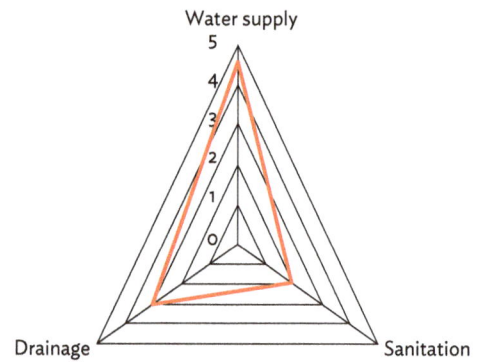

(d) Urban Water Security (KD3)

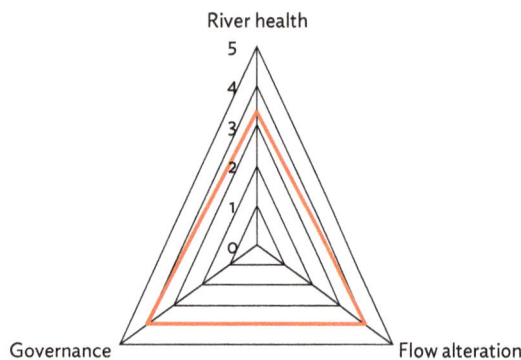

(e) Environmental Water Security (KD4)

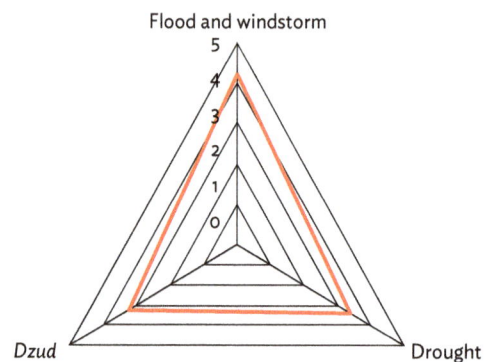

(f) Resilience to Water-Related Disasters (KD5)

KD = key dimension.

Note: *Dzud* refers to a weather phenomenon unique in Mongolia where summer drought is followed by severe winter, resulting in death of a large number of livestock.

Source: Asian Development Bank. 2017. *Country Water Security Assessment of Mongolia*. Consultant's report. Manila (TA 8855-MON).

Table 3: Water Security Scores and Core Interventions by Key Dimension

KD	Average Score (out of 20)	Score Range of the River Basins	Core Interventions to Improve Water Security
Rural household water security (KD1)	11.5	10.8–12.0	• Awareness and access to improved WSS for rural herder communities increased
Economic water security (KD2)	12.3	9.6–15.4	• Agriculture: irrigation and livestock water points expanded • Energy: renewable energy increased by 30%; water-saving measures employed for coal power plants • Mining and industry: water sources planning and management improved; effluent treatment increased
Urban water security (KD3)	11.6	7.5–13.8	• Substantial financing requirements addressed • Gaps in service level between central and *ger* areas narrowed • Levels of O&M cost recovery improved to promote sustainability
Environmental water security (KD4)	16.0	12.0–20.0	• Institutional framework for the management and regulation of water resources and the environment strengthened • Sustainable abstractions of rivers and groundwater guaranteed; monitoring and control improved
Resilience to water-related disasters (KD5)	14.1	12.0–16.7	• Investigation of various disaster risks improved to increase cost-effectiveness of interventions—key areas include drought, flood, and *dzud*

KD = key dimension, O&M = operation and maintenance, WSS = water supply and sanitation.

Notes:

1. *Ger* areas refer to Mongolia's traditional tent communities.
2. *Dzud* refers to a weather phenomenon unique in Mongolia where summer drought is followed by severe winter, resulting in death of a large number of livestock.

Source: Asian Development Bank. 2017. *Country Water Security Assessment of Mongolia*. Consultant's report. Manila (TA 8855-MON).

4.3 Water Security at the River Basin Level

The water security scores for each of Mongolia's 29 river basins have been estimated, as well as the average (both weighted and unweighted) river basin water security scores (Figure 8 and Table 4). At the river basin level, scores range from 57.2 (Khuisiin Govi–Tsetseg Lake basin) to 70.1 (Uvs Lake–Tes basin), demonstrating that there are no significant variations and no clear, distinct patterns in river basin water security across the country. Additionally, no substantial differences in water security have been noted across the three major drainage basins.

This apparent lack of variance in water security among the river basins could be traced back to the banding criteria used for some indicators: (i) a number of criteria was so stringent, resulting in all river basins getting the lowest score of 1; (ii) some criteria were adjusted to become more lenient or relaxed, allowing all river basins to obtain the maximum score; and (iii) a few others were not applicable in some river basins (e.g., hydropower and industry), resulting in a flat score for those indicators.

Based on their overall water security scores and indexes, Mongolia's 29 river basins are all categorized *capable*, with some nearing the *effective* stage (Table 1 briefly describes the water security stages and Table 4 gives the detailed river basin scores). There is ample room for improvement, particularly through the following proposed investments grouped by KD:

(i)   KD1: increased access of herders to improved WSS facilities;

(ii)   KD2: enhanced water supply to support irrigation extension, mining activities, and energy development;

(iii)   KD3: better and expanded WSS in urban centers;

(iv)   KD4: improved capacity for wastewater treatment; and

(v)   KD5: heightened resilience against water-related disasters, including development of water storage systems that collect water during floods and provide buffer against drought.

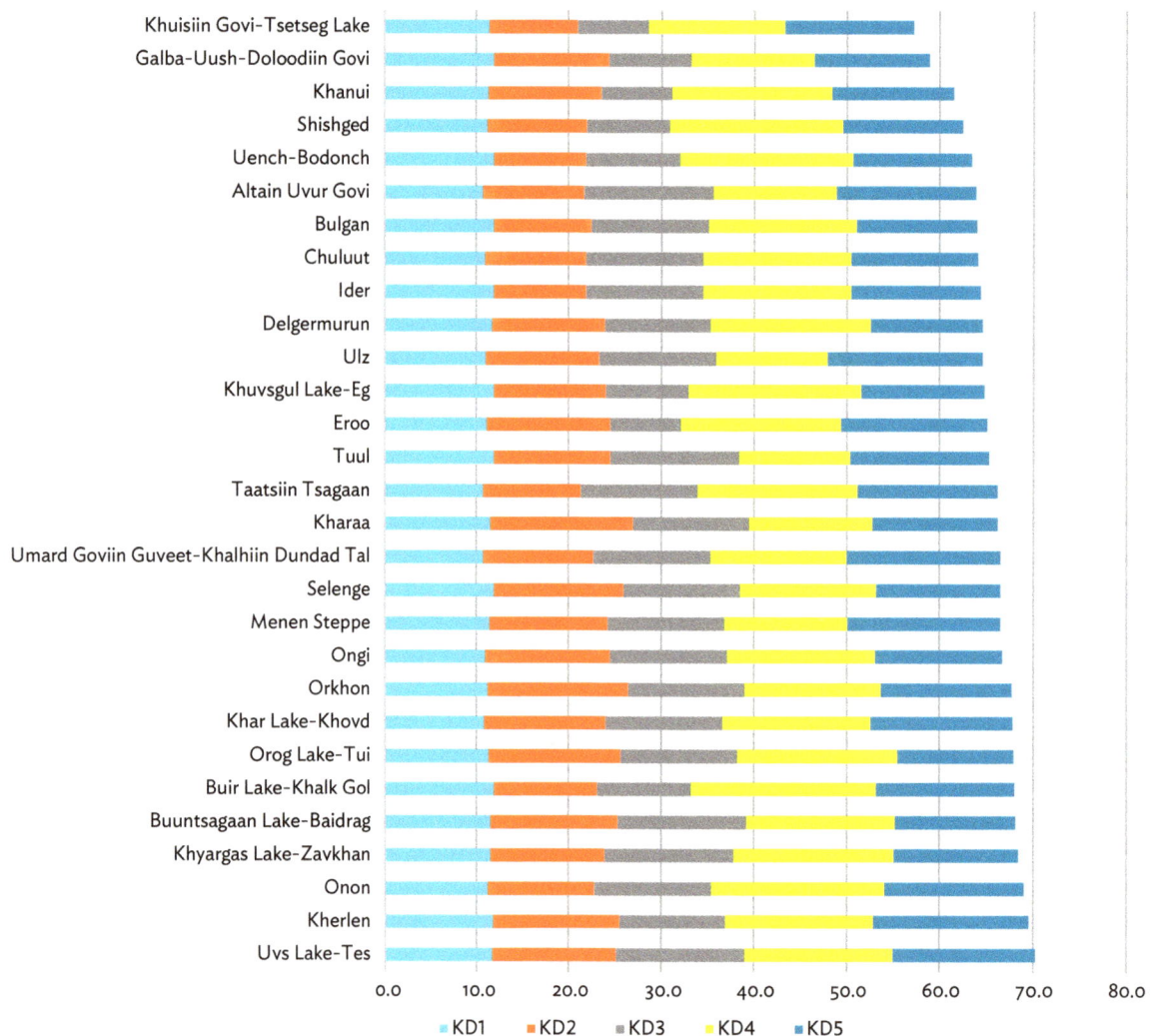

Figure 8: Mongolia's River Basin Water Security Scores

KD1 = key dimension 1 (rural household water security), KD2 = key dimension 2 (economic water security), KD3 = key dimension 3 (urban water security), KD4 = key dimension 4 (environmental water security), KD5 = key dimension 5 (resilience to water-related disasters).

Note: Each river basin water security score ranges from 1 (worst) to 100 (best).

Source: Asian Development Bank. 2017. *Country Water Security Assessment of Mongolia*. Consultant's report. Manila (TA 8855-MON).

Table 4: Water Security Scores and Indexes by River Basin

	River Basin	KD1	KD2	KD3	KD4	KD5	RBWS Score	RBWS Index
1	Selenge	12.0	14.0	12.5	14.7	13.3	66.5	3.5
2	Khuvsgul Lake–Eg	12.0	12.1	8.8	18.7	13.2	64.7	3.4
3	Shishged	11.3	10.8	8.8	18.7	12.9	62.4	3.3
4	Delgermurun	11.8	12.2	11.3	17.3	12.0	64.6	3.4
5	Ider	12.0	10.0	12.5	16.0	13.9	64.4	3.4
6	Chuluut	11.0	11.0	12.5	16.0	13.6	64.1	3.4
7	Khanui	11.4	12.2	7.5	17.3	13.1	61.5	3.3
8	Orkhon	11.3	15.2	12.5	14.7	14.0	67.7	3.6
9	Tuul	12.0	12.6	13.8	12.0	14.9	65.3	3.5
10	Kharaa	11.6	15.4	12.5	13.3	13.4	66.3	3.5
11	Eroo	11.2	13.4	7.5	17.3	15.7	65.1	3.5
12	Onon	11.3	11.6	12.5	18.7	14.9	68.9	3.6
13	Ulz	11.1	12.3	12.5	12.0	16.7	64.7	3.4
14	Kherlen	11.9	13.7	11.3	16.0	16.6	69.4	3.7
15	Buir Lake–Khalk Gol	12.0	11.2	10.0	20.0	14.8	68.0	3.6
16	Menen Steppe	11.5	12.8	12.5	13.3	16.4	66.6	3.5
17	Umard Goviin Guveet–Khalhiin Dundad Tal	10.8	12.0	12.5	14.7	16.5	66.4	3.5
18	Galba–Uush–Doloodiin Govi	12.0	12.4	8.8	13.3	12.4	58.9	3.1
19	Ongi	11.0	13.6	12.5	16.0	13.6	66.7	3.5
20	Altain Uvur Govi	10.8	11.0	13.8	13.3	15.0	63.9	3.4
21	Taatsiin Tsagaan	10.8	10.6	12.5	17.3	15.0	66.2	3.5
22	Orog Lake–Tui	11.4	14.3	12.5	17.3	12.4	68.0	3.6
23	Buuntsagaan Lake–Baidrag	11.6	13.8	13.8	16.0	12.9	68.1	3.6
24	Khyargas Lake–Zavkhan	11.6	12.4	13.8	17.3	13.3	68.4	3.6
25	Khuisiin Govi–Tsetseg Lake	11.5	9.6	7.5	14.7	13.9	57.2	3.1
26	Uench–Bodonch	12.0	10.0	10.0	18.7	12.7	63.4	3.4
27	Bulgan	12.0	10.6	12.5	16.0	12.9	64.0	3.4
28	Khar Lake–Khovd	10.9	13.2	12.5	16.0	15.2	67.8	3.6
29	Uvs Lake–Tes	11.8	13.4	13.8	16.0	15.2	70.1	3.7

continued on next page

Table 4 *continued*

	KD1	KD2	KD3	KD4	KD5	RBWS Score	RBWS Index
Average (Mongolia)	11.5	12.3	11.6	16.0	14.1	65.5	3.5
Weighted average (Mongolia)	11.4	13.0	13.1	15.2	14.6	67.4	3.6
Northern Arctic Ocean Basin	11.7	13.1	13.4	15.6	14.5	68.3	3.6
Pacific Ocean Basin	11.8	13.4	11.4	16.1	16.4	69.2	3.7
Central Asian Internal Basin	11.2	12.6	12.5	15.0	14.6	65.8	3.5
Standard deviation	0.43	1.53	2.02	2.09	1.37	2.96	0.15

lowest score   highest score

KD1 = key dimension 1 (rural household water security), KD2 = key dimension 2 (economic water security), KD3 = key dimension 3 (urban water security), KD4 = key dimension 4 (environmental water security), KD5 = key dimension 5 (resilience to water-related disasters), RBWS = river basin water security.

Notes:

1. Each key dimension score ranges from 1 (worst) to 20 (best); the overall RBWS score ranges from 1 (worst) to 100 (best); the RBWS index ranges from 1 (worst) to 5 (best).

2. Taking into consideration the uneven distribution of Mongolia's population across the country, a national score was calculated using the weighted average of all river basin-specific key dimension scores based on (i) population (rural population for KD1, urban population for KD3, and total population for KD2 and KD5) and (ii) land area (for KD4).

Source: Asian Development Bank. 2017. *Country Water Security Assessment of Mongolia*. Consultant's report. Manila (TA 8855-MON).

5 Institutional Assessment

5.1 Key Institutional and Policy Issues for Water Security

The Asian Water Development Outlook (AWDO) methodology for water security assessment, in general, and as adapted for Mongolia's CWSA (Chapter 4), provides a foundation for evaluating water security and analyzing the effectiveness of water resources planning, service delivery, and investments. Table 5 presents the linkages between the key dimensions (KDs) of water security and the water sector institutions in Mongolia.

Table 5: Water Security Key Dimensions and Associated Water Sector Institutions

KD	KD Score (out of 20)	Water Sector	Key Institutional and Policy Issues	Responsible Agencies
Rural household water security (KD1)	11.5	Rural water supply for herder communities	• Lack of WSS policy for herder households • Lack of rural WSS awareness • Low interest in WSS investment	MOFALI
Economic water security (KD2)	12.3	Irrigation	• Lack of financing and organization for efficient O&M	MOFALI
		Livestock water supply	• Impacts of water points and overgrazing on the sustainability of pastures • Poor financing and weak system to maintain existing wells	MOFALI
		Industry and mining	• Lack of coordination between RBOs, national and local governments, line ministries, and private investors • Inconsistent regulations, standards, and procedures; delays in issuance of permits; and lack of long-term planning • Uncontrolled abstractions and discharge of untreated or poorly treated wastewater to the environment • Stakeholder challenges, including local community resistance, irresponsible activities of mining companies, and conflicts of interest	MM, MOFALI, MET

continued on next page

Table 5 *continued*

KD	KD Score (out of 20)	Water Sector	Key Institutional and Policy Issues	Responsible Agencies
Urban water security (KD3)	11.6	Urban water supply and sanitation	• Insufficient tariffs and weak cost recovery to meet O&M outlays • Imbalanced WSS service levels in urban areas; lack of policy and legal agreement on provision of urban WSS service • Low investment returns • Poor performance of water service providers • Discharge of untreated or insufficiently treated wastewater to the environment	MCUD, MUB, Ulaanbaatar City governor, *aimag* (provincial) and *soum* (district) governments
Environmental water security (KD4)	16.0	Environmental water management	• Complex environmental laws, regulations, and standards • Weak enforcement of environmental policies by regulatory and implementing bodies • Lack of monitoring mechanisms to regularly check water use and effluent discharges; limited capacity to penalize noncompliance	Crosscutting MET and line agencies
Resilience to water-related disasters (KD5)	14.1	Water-related disaster management	• Lack of understanding of resultant risks from inefficient use of scarce financing • Lack of resources and training to better support disaster risk management • Limited interagency coordination	Crosscutting NEMA, MET, and other line agencies

KD = key dimension; MCUD = Ministry of Construction and Urban Development; MET = Ministry of Environment and Tourism; MM = Ministry of Mining; MOFALI = Ministry of Food, Agriculture and Light Industry; MUB = Municipality of Ulaanbaatar; NEMA = National Emergency Management Agency; O&M = operation and maintenance; RBO = river basin organization; WSS = water supply and sanitation.

Source: Asian Development Bank. 2017. *Country Water Security Assessment of Mongolia*. Consultant's report. Manila (TA 8855-MON).

5.2 Benchmarking of River Basin Organizations

River basin organizations (RBOs) have the primary responsibility of implementing water management effectively. Their two core functions involve (i) preparing river basin management plans for each river basin, and (ii) supporting the implementation of these plans. Assessment of RBO performance is done through benchmarking—i.e., evaluating how the RBOs measure up against key performance indicators in the fulfillment of their mandates. Besides baseline benchmarking, periodic benchmarking is proposed (e.g., every 5 years) to assess how the RBOs perform and to make necessary adjustments or revisions.

The methodology developed by the Network of Asian River Basin Organizations is a frequently used benchmarking approach in Asia.[26] It provides a framework of performance indicators that are comprehensive, yet flexible, to support various efforts for strengthening RBO operations. This framework has been adjusted to take into consideration the specific circumstances in Mongolia and to make it more user-friendly, but the fundamental structure and procedure of the methodology have been maintained to promote sufficient compliance with international best practice. The modified framework employs a simple semi-quantitative scoring system to facilitate a quick assessment process, which can be regularly repeated to assess the progress of integrated water resources

[26] Network of Asian River Basin Organizations. 2018. The International Workshop on RBO Performance Benchmarking. Bangkok. 20 July..

management (IWRM); this can be linked with the water security indexes. The assessment process was designed to include both river basin authorities (RBAs) and river basin councils (RBCs), and to assess their combined functions.

Baseline benchmarking is principally a self-assessment process undertaken by the RBA managers, in close collaboration with representatives of the RBCs. These managers and representatives are the most qualified to evaluate their respective organizations, including resource capability and adequacy. There were about four to six personnel from each RBO who participated in the benchmarking process—all from the RBAs and none from the RBCs. The participants demonstrated their professionalism and pragmatism as evaluators, and they were optimistic about the approach given their good grasp of its objectives.

The benchmarking, which was carried out in 2016, is grouped under five performance areas (mission, stakeholders, learning and growth, technical capacities, and finance) and measured by 14 performance indicators. Table 6 and Figure 9 present the benchmarking results for the 21 RBOs, wherein each performance indicator is rated from 0 (*worst*) to 4 (*best*). Table 7 defines the benchmarking scores based on the RBOs' stages of development.

The average overall score for 2016 is 1.8 out of a potential rating of 4.0 (Table 6), which reasonably reflects the current stage of development of the RBOs (Table 7). The RBOs have set a target score of 3.2 by end of 2020. This will be challenging to achieve and requires a significant shift from being relatively low-profile organizations to becoming drivers of IWRM. The real benefits of RBOs supporting sustainable IWRM will be evident when benchmarking scores reach at least 3.0.

Table 6: Benchmarking Performance Indicators and Scores

Performance Area	Performance Indicator	2016 Rating (0–4)		2020 Target Rating (0–4)	
		By Indicator	Average by Performance Area	By Indicator	Average by Performance Area
A. Mission	1. Status of the RBO	1.8	2.0	3.1	3.2
	2. Adequacy of institutional framework	2.2		3.2	
B. Stakeholders	3. Stakeholder satisfaction	2.1	1.9	3.2	3.0
	4. Stakeholder feedback	2.1		3.4	
	5. Stakeholder environment	1.4		2.8	
	6. Stakeholder livelihoods	1.9		3.1	
C. Learning and growth	7. Human resources development	2.0	2.0	3.4	3.4
	8. Technical development	1.6		3.1	
	9. Organizational development	2.4		3.6	
D. Technical capacities	10. Water planning and management	1.8	1.8	3.3	3.0
	11. Water allocations	1.8		2.3	
	12. Information management	1.9		3.6	
E. Finance	13. Financial independence	1.0	1.2	2.6	2.8
	14. Financial efficiency	1.4		2.9	
Average		**1.8**		**3.2**	

RBO = river basin organization.
Note: A score or rating of 0 is *worst* and 4 is *best*.
Source: Asian Development Bank. 2017. *Country Water Security Assessment of Mongolia*. Consultant's report. Manila (TA 8855-MON).

Figure 9: Average Benchmarking Scores across 21 River Basin Organizations

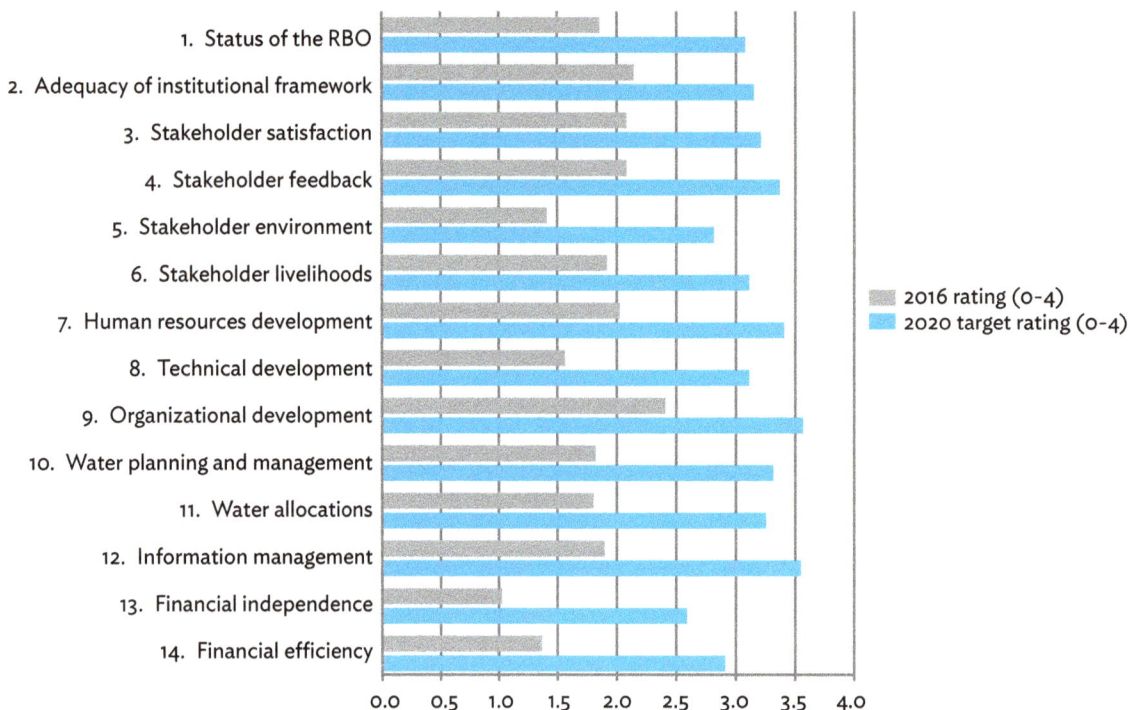

RBO = river basin organization.

Note: A score or rating of 0 is *worst* and 4 is *best*.

Source: Asian Development Bank. 2017. *Country Water Security Assessment of Mongolia*. Consultant's report. Manila (TA 8855-MON).

Table 7: Definition of Benchmarking Scores

Score	Stage of Development
0–1	• RBOs are at a preliminary stage, with facilities still being established. • There is minimal contact with and limited influence from stakeholders.
1–2	• RBOs are established and are already engaged in river basin plan preparation, data collection activities, and provision of fundamental support for water management. • Stakeholder involvement is at an early stage. • Lack of financial resources is constraining activities.
2–3	• With adequate financing in place, RBOs are making significant progress toward IWRM. • Implementation of plans is ongoing, with reasonable stakeholder participation. • Strengthened linkages between line ministries and governments (district, provincial, and central) are being developed.
3–4	• RBOs are contributing significant impact on water management through proactive and sustainable IWRM of groundwater and surface water. • Stakeholder participation is at a high level and includes active engagement with line ministries and provincial governments. • RBO financing is adequate and sustainable.

IWRM = integrated water resources management, RBO = river basin organization.

Source: Network of Asian River Basin Organizations. 2018. The International Workshop on RBO Performance Benchmarking. Bangkok. 20 July.

The assessment template also provides evaluators with a section where they can describe the current situation and summarize recommended initiatives to be carried out before the next benchmarking (proposed in December 2020). There were no inputs from the RBCs for the benchmarking, which was a drawback given that RBCs were integral elements of the original design of decentralized and participative river basin management. However, this does not invalidate the results of the benchmarking process.

A further issue, which was not assessed in the benchmarking and which is likely to become increasingly important, is the consideration of transboundary issues in water management in those river basins forming part of international river systems. For the effective management of transboundary resources, a high-level team is needed to deal with diplomatic arrangements and handle the complex and integrated management of rivers, groundwater, and the environment. While this is a defined responsibility of the Ministry of Environment and Tourism, it lacks the skills and resources to field such a team. Additionally, suitable information is not provided to RBOs, as there are constraints in consolidating water use data for the major rivers and groundwater systems.

5.3 Performance of the Service-Providing Sector

Water service providers (WSPs) manage water and wastewater on behalf of the government for the local communities. While the Ministry of Construction and Urban Development has overall responsibility for the planning and development of urban services, the urban authorities at different levels are responsible for the operation and provision of these services. The operations of urban water and wastewater services are generally commissioned to public utility service organizations, which are semi-privatized organizations working on contracts based on operation and maintenance (O&M) tariffs.

The national government may have ownership of water infrastructure, but local administrations raise the funds needed to cover most of the capital, O&M, and repair costs. Hence, *aimag* and *soum* governors can greatly influence the operations of WSPs. High dependency of *aimags* and *soums* on central government support persists because the majority of water tariffs are about half the level required to meet the full cost recovery of operations. It has been difficult for WSPs to secure tariffs adequate for cost recovery. For this reason, several public utility service organizations have resorted to cross-subsidizing the shortfall in revenue from water supply with income from more financially viable sectors, such as energy supply. In some *soums*, the task of providing support to the management of water supplies has been assigned to local community organizations.

5.3.1 Weaknesses of Water Service Providers

Inadequate O&M of water sector investments has resulted in the depreciation and poor performance of many water infrastructures, with some investments failing to meet their targeted output, thereby adversely affecting water security. The capacity and financial resources for the effective management of water schemes are sorely lacking among government-owned and community-based WSPs, as shown in the following examples:

(i)   urban WSPs are unable to meet the requirements of an expanding urban population;

(ii)   targets in the management contracts bear little relation to performance problems in the system;

(iii)   water supply in *aimag* and *soum* centers include partial supply of piped water, and the demand of local populations for sewerage services is low since affordability is a problem;

(iv)   coordination among agencies and different levels of government is lacking, and interaction with stakeholders is low;

(v) low economic and financial returns, alongside low cost-recovery levels, constrain investments in water utilities; and

(vi) use of public–private partnership schemes in the water sector (e.g., build–operate arrangements) has been sluggish.

5.3.2 Strengthening of Water Service Providers

With investments in government-owned infrastructure comprising a major part of the required total investments, robust and self-sufficient WSPs are vital components for successful IWRM implementation. To achieve this, WSPs should have clearly defined performance targets and the accountability to meet them. There should be adequate financial incentives for meeting performance targets and/or penalties (disincentives) if targets are not met. Greater autonomy, including flexibility in management and tariff setting, will help WSPs meet their targets. Instead of a fixed rate for all, differential rates for different classes of users may be the way to go. Households may pay a low tariff, but hotels, industries, and other commercial users may be required to pay a higher tariff. Differential rates may likewise be imposed for wastewater services, with industries paying higher charges to reflect the costs of treating industrial discharges. It is also important to check how financial returns can be balanced with social objectives. Further recommendations on strengthening the WSPs are given in Section 6.2.2.

5.4 Summary of the Institutional Review

Table 8 summarizes Mongolia's institutions and water governance, following the assessment framework and core conditions for IWRM as described by the United Nations Educational, Scientific and Cultural Organization.[27]

Table 8: Summary Findings of the Institutional Review

Core Condition for IWRM	Preliminary Assessment
Political will and commitment	• Strong official commitment to IWRM and the promulgation of the 2012 Water Law provide a sound basis for efficient IWRM and decentralization. However, water resources planning and investments are generally sector-based, and decision-making is still centralized. • There remains a substantial communication gap between the line ministries, local governments, and RBOs. There is also the challenge of RBOs having to deal with several *aimag* (provincial) governments within their river basin, just as *aimag* governments have to deal with several RBOs. • Developing delegation and good working procedures takes time and needs to be supported.
Basin development plan and clear vision	• River basin plans have been prepared or are under preparation by RBAs; some have been approved by local citizens' *khurals* (parliaments). Some plans are preliminary and need to be updated and made more specific (i.e., operational). • RBAs need access to specialist skills for this.
Participation and coordination mechanisms	• The establishment and operation of the RBCs lag behind. Theoretically, having the RBAs and RBCs assume parallel roles is a good approach, but it is not clear how they should each execute their roles. • RBOs need to focus on integrated river basin functions and should avoid duplicating tasks undertaken by other agencies. *Aimag* and *soum* (district) governments could assume some of the RBO roles.

continued on next page

---

27 United Nations Educational, Scientific and Cultural Organization, International Hydrological Programme; and Network of Asian River Basin Organizations. 2009. *IWRM Guidelines at River Basin Level.* Paris.

Table 8 *continued*

Core Condition for IWRM	Preliminary Assessment
Capacity development	• The RBOs are small, having fewer than 10 employees per unit. MET has made good progress in supporting capacity development. • It is important for RBOs to be able to engage specialists, as and when needed; however, RBOs have problems accessing and maintaining high levels of expertise.
Investment and policy recommendations	• Various agencies are involved in investment and policy decision-making, but RBOs sit outside the decision-making hub. • With many RBCs being newly established (and others still being established), they are largely not involved in decision-making. Also, support to strengthening their role is limited.
Well-defined and flexible legal frameworks and regulation	• Implementation of water resources legislation mostly depends on the *aimag* and *soum* governments; but, with unclear mandates, deficient regulation, and weak enforcement capability, they have limited scope to execute water resources plans and water licensing requirements at local levels.[a] • Some responsibilities proposed for decentralization remain with the central government, thereby creating ambiguity and overlapping of some functions.
Water allocation plans	• The level of interaction between line agencies and the RBOs needs to be strengthened. • Water investments and management are under the central government as well as the *aimag* and *soum* administrations; the decentralized RBOs lack resources and capacity to engage with central water sector agencies. • There is a need to clearly define the responsibilities regarding licensing and monitoring of water use and effluent quality. RBOs should be strengthened and should take a more proactive role.
Adequate investment financial stability and sustainable cost recovery	• External funding from donor agencies or development partners has heavily supported the new initiatives for IWRM. In the RBO benchmarking exercise, financial efficiency and financial independence earned the lowest scores; hence, opportunities for long-term and sustainable financing need to be considered. • Water service delivery is weak, mainly because water service providers are poorly funded and inept.
Good knowledge of the natural resources	• Knowledge gaps need to be closed, especially relating to groundwater. • The decentralized river basin management leaves weaknesses in coordinated water resources management, and there is a lack of access to specialist skills.
Comprehensive monitoring and evaluation	• Water use and hydrological data surveys are undertaken by MET and RBOs. Better targeted and cost-effective surveys are needed, which are linked to mechanisms for assessment and proactive initiatives. Groundwater monitoring and analysis are weak. • Information from different organizations needs to be compiled.

IWRM = integrated water resources management, MET = Ministry of Environment and Tourism, RBA = river basin authority, RBC = river basin council, RBO = river basin organization.

[a] P. Tortell, A. T. Borjigdkhan, and E. Naidansuren. 2008. *Institutional Structures for Environmental Management in Mongolia.* Ulaanbaatar and New York: United Nations Development Programme.

Source: Asian Development Bank. 2017. *Country Water Security Assessment of Mongolia.* Consultant's report. Manila (TA 8855-MON).

6 Planning for Action

B ased on the analysis carried out in the 2017 country water security assessment (CWSA) of Mongolia on the five key dimensions (KDs) of water security, an investment program has been developed that sets out the direction and lists the key water sector investments for 2018–2030 (footnote 4). The proposed investment program is worth $6.50 billion, to be financed as follows: $3.67 billion (56%) from the private sector, $1.76 billion (27%) from the government, and $1.07 billion (16%) from the beneficiaries. The program includes requirements by the government, the private sector, and beneficiaries for investments in water-related infrastructure, in conjunction with initiatives to effectively manage and ensure the sustainability of water resources. The investment program builds on the national Integrated Water Management Plan (IWMP) of 2013 (footnote 5) and follows the established targets of Mongolia's Sustainable Development Vision (SDV) 2030 (footnote 17). It is designed to integrate the targets of the SDV 2030 with the goal of strengthening water security alongside broader government policies and objectives to strengthen economic diversification, promote employment creation, and support environmentally sustainable development. The investment program outlines physical investments, as well as the key requirements for technical assistance to support these investments. In addition, recommendations are made on actions to strengthen the institutions in the country, particularly those related to water.

6.1 Outline of the Physical Investments

6.1.1 Investments to Achieve Water Security in the Five Key Dimensions

The core areas of investment in the water supply and sanitation (WSS) sector, assessed within the KDs of water security are as follows:

(i) **Investments in rural household water security (KD1).** These target improved WSS for rural communities in accordance with the SDV 2030. The program will raise awareness of the benefits of improved WSS and the risks of poor WSS. Further, it will provide financial support to herders and their families to enable them to invest in low-cost WSS.

(ii) **Investments to support economic water security (KD2).** These will promote (a) adequate water supply and water quality to support economic activities; and (b) an enabling environment for investments that meet government objectives, including sustainable and environmentally sound water usage, economic diversification, employment generation, reduction in energy and water use, and promotion of renewable energy. Core areas of physical investments include (a) development of 70,000 hectares of new and upgraded irrigation area; (b) livestock and pasture management initiatives, including ventures in livestock water points; (c) projects for improved and sustainable water supplies in the South Gobi region to support economic growth; and (d) construction of 650 megawatts of hydropower.

(iii) **Investments in urban water security (KD3).** These follow the SDV 2030 targets and are designed to support the development of improved water services for Mongolia's rapidly expanding urban centers to address water demand for domestic and industrial use as well as for municipal and public services. Investments are targeted to maximize the coverage of improved WSS for urban households, enhance effectiveness of wastewater treatment, improve water for industry, and support economic growth in urban and peri-urban areas. Investments will be designed to be financially and economically viable and ensure sustainable levels of revenue from operational costs. Physical investments will improve WSS facilities, supplies to industry, and municipal and public services in Ulaanbaatar, 21 *aimag* centers, and 330 *soum* centers.

(iv) **Investments in environmental water security (KD4).** These include ensuring policy and institutional frameworks are in place to allow for sustainable water resources management, with environment-related investments made through the various water sector programs. Environment-specific and crosscutting investments are proposed to support water resources protection and enhancement, such as wastewater treatment, protection from mining spoils and tailings, erosion control, conservation of riparian vegetation, and responsible increase of agrochemical use to avoid nonpoint source pollution from agriculture.

(v) **Investments to reducing water-related disasters (KD5).** These focus on knowledge enhancement and awareness raising about risks from disasters to make investments in drought, flood, and *dzud* protection more effective; early warning for droughts and spring malt floods; and disaster management and awareness raising to reduce water-related disaster risk.

The investment program supports the reduction of emissions that are driving climate change by promoting increased energy and water use efficiency for power production, as well as by meeting 30% of total energy demand with renewable energy sources. Some increase of energy use can be expected from pumping water for the expansion of industry, mining, urban water supply, and irrigation.

The physical investment will be supported and funded by the Government of Mongolia and, in some cases, by the private sector (where private financing is recommended). Details on the proposed investments are described in the CWSA report (footnote 4).

6.1.2 Technical Assistance

The key areas of technical assistance are as follows:

(i) **Planning and design of investment programs.** To assess the economic, financial, and technical viability of investments, they should be supported by planning, detailed design, and feasibility studies. Due diligence is essential in evaluating financing alternatives, developing logical and low-cost solutions, and maximizing economic returns. Other project preparatory studies, such as environmental and social impact assessments, can help promote environmental and social benefits (e.g., ecosystem preservation, poverty alleviation, health protection, post-COVID-19 response, and employment generation).

(ii) **Stakeholder engagement, capacity building, institutional development, and policy support initiatives.** Investments should be backed by strong initiatives of stakeholder participation, training, and institutional development. Technical assistance can be tapped in identifying areas requiring urgent policy and regulatory adjustments.

(iii) **Improving knowledge of water resources.** It is important to prepare detailed and location-specific assessments of surface water and groundwater, including assessments on climate change impacts. Studies that further advance the understanding of groundwater resources, especially those in the vicinity of the South Gobi region's mining and industrial areas, are needed.

(iv) **Strengthening integrated water resources management.** To enhance the sustainable management of water resources, including integrated water resources management (IWRM), the capacity of key agencies such as river basin organizations (RBOs) should be strengthened. Coordination between agencies and stakeholders should also be promoted, along with strengthening of water policies and regulatory frameworks. Moreover, crosscutting water resources assessments are required.

(v) **Strengthening of private sector partnerships.** A significant portion of the proposed investment program will be undertaken through partnerships with the private sector. Strengthening private sector engagement enables the provision of substantial financial and operational support through various schemes, including public–private partnership arrangements.

6.1.3 Hydro-Economic Assessments

Hydro-economic assessments have been carried out for several investment types. A hydro-economic assessment expresses the incremental cost of providing additional water supply per unit volume of water (e.g., United States dollars per cubic meter). It incorporates both investment and operational costs using financial discounting.

Figure 10 illustrates the cost curves prepared for the four main areas of incremental water demand: the capital city of Ulaanbaatar, *aimag* and *soum* urban centers, irrigation, and mining.

Figure 10: Cost Curves for the Main Areas of Incremental Water Demand, 2017
($ per cubic meter)

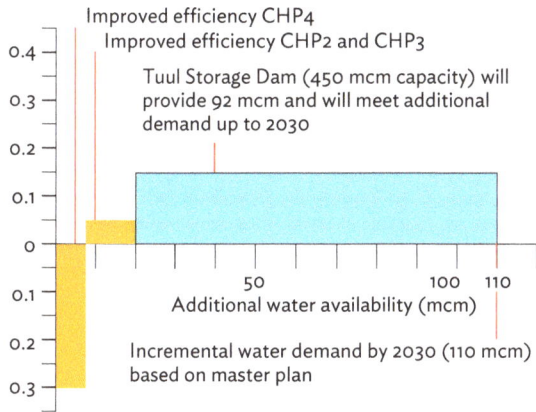

(a) Ulaanbaatar Water Supply

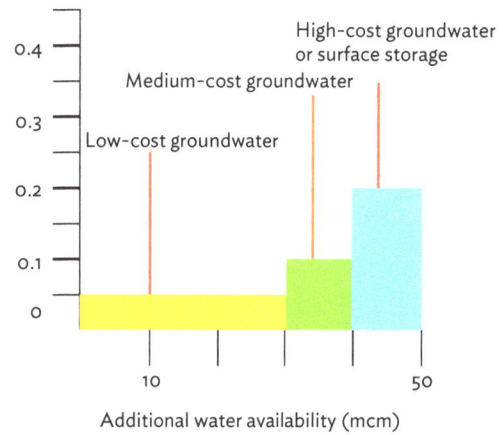

(b) Urban Water Supply for 21 *Aimag* Centers and 330 *Soum* Centers

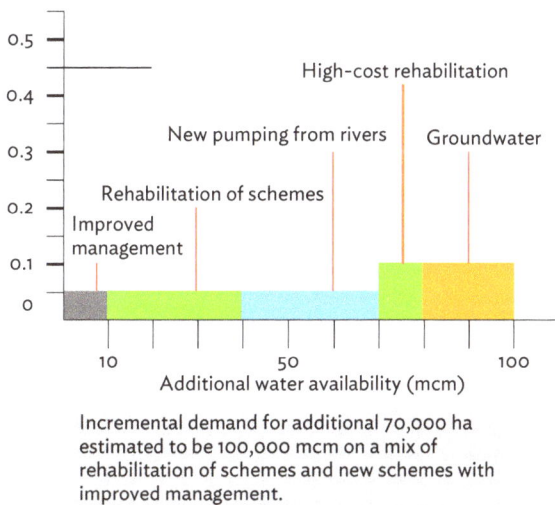

(c) Irrigation

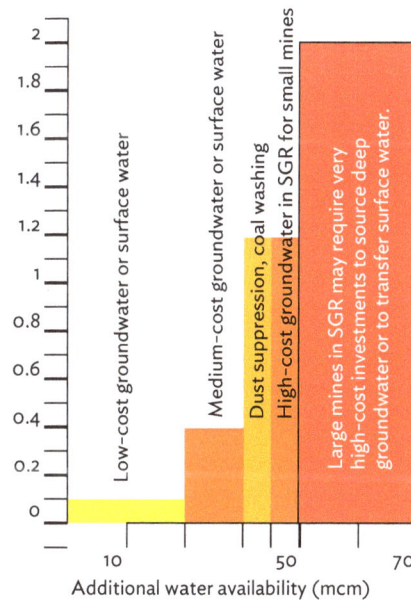

(d) Mining

CHP = combined heat and power, ha = hectare, mcm = million cubic meter, SGR = South Gobi region.

Notes:

1. The cost curves are expressed in cost of additional water based on cost-effectiveness ratio ($ per cubic meter).
2. *Aimag* refers to the provincial administrative unit in Mongolia, whereas *soum* is the subprovincial administrative unit.

Source: Asian Development Bank. 2017. *Country Water Security Assessment of Mongolia*. Consultant's report. Manila (TA 8855-MON).

6.2 Strengthening Institutions

6.2.1 Implementing Integrated Water Resources Management

There is a strong commitment to IWRM, as defined in the 2012 Water Law; and, since 2012, good progress has been made to strengthen water management. To date, 21 river basin authorities (RBAs) have been established, with basic capacity to monitor water use and contribute to decision-making. Progress lags in the establishment of river basin councils (RBCs); nonetheless, the creation and operationalization of RBCs are critical.

The next stage in the development of decentralized IWRM is to push the RBOs from their current supporting and monitoring roles into becoming drivers of sustainable IWRM. However, this is a challenging step. The institutional analysis described in Chapter 5 identifies core weaknesses in the RBAs (including lack of budget) and weak linkages with the *aimag* and *soum* governments and the other water sector agencies.

River basin IWRM plans are being prepared for each of the 29 river basins. The plans describe how the water is to be managed and developed. They also indicate how water should be allocated between different sectors. So far, only 17 of the 29 river basin plans have been completed and endorsed. The following recommendations are made to strengthen the institutional framework:

(i)   There is a need for various policies and regulations to be coherent. The approximately 20 organizations involved in water resources need to have clearly specified roles that are complementary. It is likewise important to prepare an integrated water policy document covering all water uses.

(ii)   The linkages between the line ministries, *aimag* and *soum* governments, and the RBOs should be clarified and reinforced. The central government, through the National Water Committee, needs to provide clearer guidance on procedures, tariffs, and licensing processes.

(iii)   Preparation and implementation of investment projects are sluggish and protracted. The project cycle of planning, feasibility studies, detailed design, tendering, implementation, and monitoring should be formally agreed on and improved. Approval procedures should also be defined and streamlined.

(iv)   Financing of the RBOs should be reviewed. Budgets need to be increased, while tasks need to be consolidated to reduce redundancies with other agencies and, thereby, increase efficiencies.

(v)   The role of the RBOs should be strengthened and expanded in the next stage of their development. This can be achieved through (a) review and clarification of the existing legal and policy framework; (b) provision of increased support through higher-level bodies, including the National Water Committee; (c) establishment of effective RBCs; and (d) enhancement of capacities to effectively support water management.

(vi)   The institutional framework for implementing the river basin plans needs to be clarified. Since implementation of plans involves multiple institutions, it is vital for the roles and responsibilities of the ministries, a*imag* and *soum* governments, and the RBOs to be better defined while a stronger working relationship is developed.

(vii)   The lack of clear guidance on the process of decentralization results in duplication of functions and an increased bureaucracy in decision-making. A balance should be attained, and the level of devolved responsibilities clearly set out.

(viii) The RBAs are small administrative units and sit outside the main decision-making matrix of the central, *aimag,* and *soum* governments. There are practical difficulties of maintaining many small decentralized river basin units. RBAs have to coordinate with a number of *aimags* and, in the same way, *aimag* administrations have to liaise with several RBAs. This reduces the effectiveness of the RBOs to support control and management functions with local governments.

6.2.2 Strengthening of Water Service Providers

Water service providers (WSPs) should be strengthened to facilitate the operations of government-owned water investments. This is critical to support the investment program and to help bridge gaps in government finance and capacities. Management contracts for urban WSS are weak and have significant shortcomings in achieving efficiencies.[28] New approaches are required to bolster the effectiveness of WSPs, including community-based management systems.

The WSPs should have clear performance targets, and they should also be made accountable to meet these targets, both to the government and to their customers. In return, WSPs should have adequate financial incentives, or disincentives (penalties) if targets are not met. They need greater autonomy, including flexibilities in management and support, to better engage with their customers. Experience has shown that WSPs work best within a commercial framework, balancing investments with revenues. Given the monopolistic nature of the service provision, a strong and independent regulatory body, which reports publicly, is needed to ensure compliance with management, financial, and technical policies and procedures. Such new approaches should be pilot tested in Ulaanbaatar, where economic factors are more favorable and where developing management processes can be given proper oversight.

Water tariffs. To be sustainable, water tariffs should be set at a level adequate to cover operation and maintenance (O&M) costs. In Ulaanbaatar, current tariffs are 10% lower than the O&M costs. The smaller urban centers will likely have higher costs yet a lower revenue base than Ulaanbaatar. There is a need to enhance the institutional capacities to assess and manage the collection of sustainable tariffs. Major inequities exist in the level of service for urban WSS between the central areas and the peri-urban areas—e.g., ger residents are charged higher tariffs for water supply that is of lower level of service than apartment residents.

6.2.3 Implementing Investment Programs

To successfully implement the water sector investment program, capacity and resources for planning and management should be strengthened, including access to specialist expertise for location-specific technical, legal, and environmental needs. An increased understanding of groundwater is important.

Investments in the water-related sectors, including urban WSS and agriculture, are hampered by poor financial and economic returns and by lack of funding for O&M. There are also many uncertainties surrounding the availability of government finance for the water sector, which is heavily dependent on the performance of the commodities sectors; hence, there is a need to improve the self-sufficiency of water resources investments.

The important role of the private sector in supporting water resources development is recognized. However, capacities and skills should be honed in the most effective ways to allow government and the private sector to work more closely together. The development of mines and industries is similarly based on private sector investment. Therefore, new initiatives to promote mining investments should be developed in parallel with the introduction of controls and technologies to ensure conservation of water and support environmentally sound approaches of managing effluents.

28 ADB. 2011. *Technical Assistance to Mongolia for Ulaanbaatar Urban Services and Ger Areas Development Investment Program.* Manila (TA 7970-MON).

6.2.4 Adjustments to the Legal and Policy Framework

The legal and regulatory framework is largely in place. There is, however, a need to review and simplify the multitude of regulations, clarify inconsistent areas, and provide guidance on how coordination between different sectors can be strengthened. An awareness-raising campaign and/or capacity-building program may also be required, especially for staff of agencies and organizations tasked to implement the framework at all levels.

The investment program should be supported by adjustments to the legal and policy framework, with consideration of the following key points:

(i)   There are inconsistencies between different laws, which need to be analyzed and clarified, including the institutional responsibilities and mechanisms for implementation. Inconsistencies also exist in the water resources and environmental regulatory frameworks.

(ii) There are duplications of many water tasks and decision-making remains very centralized despite policies for decentralization.

(iii) The linkages and sharing of tasks between RBOs and the *aimag* and *soum* governments need to be clarified.

7 Summary of Lessons Learned and Recommendations

The transition of Mongolia in the early 1990s—from nearly a century of Soviet-modeled rule and economy, with a strongly hierarchical and centralized system of government, toward a market economy under a democratic government structure—has had enormous impacts on the country's water management system. Basic water-related services provided previously by the government deteriorated quickly and, given their low economic and financial returns, these services were not attractive to private sector investors. In addition, professional interest in water management decreased as other sectors became more appealing from a financial income point of view.

Realizing the problem, the government started to take action in the late 1990s by establishing the National Water Committee in 1999 and the Water Authority in 2005 under the then Ministry of Nature and Environment. The 2004 Water Law introduced the legal basis for integrated water resources management (IWRM), which was further strengthened in the modified Water Law of 2012. As the basic principle for Mongolia's water management, IWRM was laid down in the extensive Integrated Water Management Plan (IWMP) of Mongolia, published in 2013 and formally ratified in Parliament in the same year as the basis for all actions in the field of water management (footnote 5).

Implementing IWRM takes time. The multisector approach of IWRM requires the involvement of many institutions, and establishing effective cooperation between all these institutions is difficult. Moreover, water systems do not stop at administrative borders, which makes cooperation between *aimags* and *soums* crucial. Strengthening capacity for this cooperation between stakeholders requires investments in institutions and capacity building. Experiences in other countries show that introducing such fundamental changes in governance takes at least 20–30 years. On top of this, infrastructure investments are expensive, and available budget allows only step-by-step implementation of required investments. For all these reasons, IWRM should be seen as a process. The IWMP and the 2017 country water security assessment (CWSA) of Mongolia (footnote 4) are part of the first stage of IWRM implementation. This development needs to be continued and actively supported.

The Asian Water Development Outlook (AWDO) approach has proven to be an effective way to assess Mongolia's water security at the national and river basin levels and to make recommendations on how to further improve the situation. The main challenge is finding the means to implement the recommended actions. Given the financial constraints, government prioritization is needed and financing must be secured. The proposed actions have to be translated into bankable projects including the identification of the mode for implementation, the role of the private sector, the strategies for funding (i.e., who pays back the investment) and financing (i.e., who brings in the money for the investment), and how the project will be procured.

Special attention should be given to making the decentralized approach to IWRM by means of the river basin organizations (RBOs) more effective. This will require strengthening the institutional setting of the RBOs, enhancing their financial means, and building the capacity of river basin authority (RBA) staff and river basin council (RBC) members.

IWRM is a continuing process. Therefore, the CWSA should be made a recurrent activity to regularly evaluate Mongolia's progress in achieving water security—and develop further recommendations on actions to be taken. It is recommended that the next CWSA be published in 2022, which means assessment should commence in 2020.[29] In the next CWSA, further changes and adjustments should be considered based on the experience with the 2017 CWSA. For example, the criteria and the banding of scoring tables for the key water security KDs could be adjusted to increase variance over river basins.

The 2017 CWSA resulted in a range of recommended investments to improve water security, structured around the five KDs. Project proposals were then developed from these recommendations to match each KD. These projects should be implemented in parallel with institutional development and technical assistance support aligned to the objectives of each project.

The team that designed and prepared the Tuul River Basin Health Report Card (footnote 7) proposed regular reassessments of the health of the Tuul River Basin—as well as other river basins in Mongolia—using the report card process. Consultations with stakeholders during preparation of the Tuul River Basin Health Report generated the following recommendations to improve the health of the Tuul River Basin:

(i)   Reduce dependence on groundwater supplies through wastewater recycling at the central wastewater treatment plant of Ulaanbaatar.

(ii)   Make a plan for a sustainable water supply for Ulaanbaatar, ensuring sustainable abstraction of surface water and groundwater resources, and maintenance of dry season flows in the river near Ulaanbaatar.

(iii)   Improve monitoring, reporting, and data management of surface water and groundwater quality and quantity to support the future sustainability of water resources.

(iv)   Improve enforcement of regulations prohibiting mining and development activities in protected zones adjacent to the Tuul River and other water bodies (as specified in the 2012 Water Law).

(v)   Incentivize upgrades in water treatment and increased water efficiency of industries (e.g., leather manufacturers) to reduce soil and water pollution and curb over-abstraction.

(vi)   Encourage sustainable population growth outside of Ulaanbaatar through improved access to drinking water and sanitation (e.g., bio-latrines) in other *soum* centers and rural settlements.

(vii)   Reduce air pollution in Ulaanbaatar through government subsidies and incentives for alternative energy sources for household consumption, particularly in the *ger* areas.

(viii) Increase spending on water resources protection and restoration to 35% of the water utilization fee, as designated in the 2012 Water Law.

(ix)   Expand aquatic biological monitoring (regions V and VI), air quality monitoring (installation of air quality stations in regions I, III, IV, V, and VI), and water quality and quantity assessments of surface water (region II).[30]

---

[29]   The next AWDO is planned to be published in 2020 and will provide an update from the situation in 2016.

[30]   The Tuul River Basin was divided into six regions based on ecosystem condition, urbanization, socioeconomic development, water use, and water pollution (footnote 7, p. 3).

(x)   Limit the number of livestock to within rangeland carrying capacity and promote intensive farming in remote areas with high pasture productivity and quality.

(xi) Encourage participation of different government agencies and stakeholders in the implementation and future revisions of the 2012 Tuul River Basin Integrated Water Management Plan.[31]

Taking into account the current development activities in Mongolia, which are supported by international donors and development partners, three possible priority investment projects are described in the Appendixes through project concept notes:

(i)   integrated water supply and sanitation (WSS), water points, pasture, and livestock production systems (to benefit herder communities);

(ii)   strategic investment for WSS in *ger* areas of Ulaanbaatar; and

(iii) integrated investment for flood protection in the urban areas (by river basin).

---

[31]   Ministry of Environment and Green Development. 2012. *Tuul River Basin Integrated Water Management Plan.* Ulaanbaatar.

Appendix 1

Project Concept Notes for Possible Priority Investment Projects: Integrated Water Supply and Sanitation, Water Points, Pasture, and Livestock Production Systems

Table A1.1: Project at a Glance

Project Title	Integrated Water Supply and Sanitation, Water Points, Pasture, and Livestock Production Systems
Summary	The project would support the upgrading of water points, pasture, and livestock production systems to improve financial returns and reduce the number of livestock to sustainable levels. The project would also develop improved WSS for herder families.

A. General Information		
Area of Water Security	*Investment Sector*	*Supporting Initiatives*
1. Household	■ Herder WSS	☐ IWRM
2. Economic	☐ Irrigation ■ Livestock ☐ Mining ☐ Industry ☐ Energy	☐ Water quality ☐ Surface water ■ Groundwater ■ Agriculture ☐ Poverty ☐ Private sector PPP
3. Urban Water Supply	☐ Ulaanbaatar water supply ☐ Major urban centers ☐ *Soum* centers	
4. Environment	☐ River basin governance ☐ Overall water resources management	☐ Catchment and protected areas ☐ Water resources protection and enhancement
5. Resilience to Water-Related Disasters	☐ Urban flood protection and management ☐ Integrated disaster management	

B. Key Thematic Areas	
■ Sustainable economic growth ■ Inclusive social development ■ Governance ☐ Gender and development	■ Environmental sustainability ☐ Regional cooperation ☐ Private sector development ■ Capacity development

continued on next page

Table A1.1 *continued*

C. Coverage	

■ 40 *soums* in 9 *aimags* as below:

Aimag	Soum
Arkhangai (2 *soums*)	Tariat, Under-Ulaan
Bayankhongor (2 *soums*)	Erdenetsogt, Galuut
Bayn-Ulgii (2 *soums*)	Noggonnuur, Ulaankhus
Bulgan (2 *soums*)	Khangal, Khutag-Undur
Gobi-Altai (2 *soums*)	Biger, Tonkhil
Tuv (24 *soums*)	Altanbulag, Argalant, Arkhust, Batsumber, Bayan, Bayan-Unjuul, Bayanchandmani, Bayandelger, Bayanjargalant, Bayankhangai, Bayantsagaan, Bayantsogt, Bornuur, Delgerkhaan, Erdene, Erdenesant, Jargalant, Mungunmorit, Sergelen, Sumber, Tseel, Ugtaaltsaidam, Undurshireet, Zaamar
Khentil (2 *soums*)	Unmnedelger, Bor-Under
Dundgobi (2 *soums*)	Delgerkhangai, Erdenedalai
Sykhbaatar (2 *soums*)	Munkhkhaan, Bayandelger

D. Responsible and Supporting Units

■ Responsible ministry: Ministry of Food, Agriculture and Light Industry
■ Supporting ministry: Ministry of Environment and Tourism

E. Linkage to the Investment Program

1. Technical assistance
   (i) Planning and design for herder WSS
   (ii) Planning and design for pasture management, livestock water points, production, and marketing schemes
2. Investment
   (i) Investment in low-cost sanitation and household water treatment
   (ii) Investment in new water points, pasture, and livestock management systems in unused and existing grazing areas

IWRM = integrated water resources management, PPP = public–private partnership, WSS = water supply and sanitation.
Note: *Aimag* refers to the provincial administrative unit in Mongolia, whereas *soum* is the subprovincial administrative unit.
Source: Asian Development Bank. 2017. *Country Water Security Assessment of Mongolia*. Consultant's report. Manila (TA 8855-MON).

1.1 Project Rationale

The present livestock system, which has evolved and has been adapted by herders, is a low-input and low-output system, and is financially viable. However, there is a lack of capacity and resources to invest in inputs to improve productivity, which is affecting sustainability. Rangeland health is the primary challenge to sustainable livestock production; the number of livestock has significantly increased in recent years. The 2015 livestock census recorded an estimated 56 million heads—the largest livestock population in history and equivalent to 100 million animals per unit of sheep. Based on internationally accepted standards, the carrying capacity in many areas has been greatly exceeded. Overgrazing is cited as a principal factor causing the degradation of pastureland; thus, control of livestock numbers is a fundamental precondition for effective rangeland management.[1] Climate change assessments conclude that increased frequency of drought, coupled with continued overgrazing, will worsen

---

[1] Government of Mongolia, Ministry of Food and Agriculture; and Swiss Agency for Development and Cooperation. 2015. *National Report on the Rangeland Health of Mongolia*. Ulaanbaatar.

conditions, leaving a large area of land fragile by 2030. Government subsidies for wool production promote increased livestock numbers.

The environmental conditions to sustain the productivity and biodiversity of pastureland have been in decline in many areas. Access to water, both natural and constructed water sources, significantly impacts the number and grazing patterns of livestock and, as a result, affects the security of livestock production. The significant increase in stocking number since the open market economy is partially because of the opening of new grazing lands around water points. The Food and Agriculture Organization of the United Nations noted that the development of water sources must await both the granting of grazing rights and the establishment of effective organization of herder households before water development has a realistic chance of success.[2]

Overgrazing problems exist around traditional and constructed water sources. Hence, the development of additional boreholes needs to consider wider management issues. Analysis shows the complexity of the livestock sector and the need to balance economic returns against the risks of irrecoverable long-term degradation of pastureland by uncoordinated development of water points.

Of the estimated 128.0 million hectares of total pastureland area, about 5.0 million hectares (4%) have been rendered unusable for animal husbandry operations because of water shortage, and another 14.1 million hectares (11%) are located in areas too remote for effective livestock production.[3] Given the needs assessment of one well requirement per 5,000 hectares of grazing lands, 1,000 wells would be needed to open up 5.0 million hectares of unused pastures. In existing pasture areas, many boreholes do not function because of lack of maintenance and lack of capacity to maintain the wells and pumps, reducing the area available for grazing because of poor access to water. It is estimated that investments in rehabilitation of existing water points and construction of new water points would allow for an increased area for grazing of nearly 10% of the pasture area, or about 12.0 million hectares.

Ensuring long-term sustainability and avoiding risk of continued degradation of pastureland requires new initiatives and reforms in pasture and livestock management as well as legal, technical, and institutional support and training. The investments need to demonstrate clear financial returns to herder households through better quality but fewer livestock. The investment is designed to improve production of livestock through investment in new water points, which will open up areas with a lack of access to water points, in parallel with investments to address overgrazing problems through integrated pasture and livestock management initiatives. Investments would include support to establish herder groups, new and rehabilitated water points, improved breeding and fodder production, strengthened marketing, development of small-scale irrigation, enhanced capacities for operation and maintenance (O&M) of pumps and boreholes, and integrated support for rural livelihoods.

1.2 Objectives

The three key objectives of this proposed water-for-livestock project are (i) increased production of livestock in pasture areas by providing and/or rehabilitating livestock water points; (ii) development of an integrated package of initiatives for sustainable pasture and livestock management, designed to increase production returns and reduce stocking numbers; (iii) effective O&M of pasture water points; and (iv) improved water supply and

---

[2]   J. M. Suttie. 2000. *Country Pasture/Forage Resource Profiles: Mongolia.* Rome: Food and Agriculture Organization of the United Nations.
[3]   B. Enkhmaa and O. Naran-Ochir. 2014. *The State of the Pasture in Mongolia.* Ulaanbaatar.

sanitation (WSS) for herder families. The project would support the development of the *bag* (village) and *khotail* (group of herders) as small institutions to become the nuclei of knowledge and capacity to initiate change.

1.3 Impact, Outcome, and Outputs

The project impact will be long-term sustainable livelihoods for herder families in 40 *soums* (districts). The project outcome will be sustainable and climate-resilient pasture and livestock management in project *soums*. The project will have three outputs:

(i) **Output 1: Plans for improved rural water supply and sanitation and sustainable livestock management are developed.** Under this output, integrated approaches to support long-term sustainable pasture and livestock production and management will be researched and developed, including technical, social, and economic studies supported by extensive consultation with herder communities. Studies will analyze the optimum location of water points based on an integrated plan for pasture management to ensure sustainable production and support of rural WSS. This would involve significant use of geographic information system and the introduction of proven pasture management software systems. The project to improve pasture, production, and marketing will be planned and designed to cover the following areas: management of pastures and grasslands, improved breeding stock and animal health, small-scale irrigation from water points, increased availability of fodder, enhanced opportunities for rural employment, and strengthened marketing systems. Extensive consultation, training programs, and awareness-raising activities with herder communities will be implemented to support the planning process.

(ii) **Output 2: Rural water supply and sanitation and livestock water points are upgraded and sustainable livestock management is in place.** This output will support an integrated program covering rural WSS and pasture and livestock management over an area of 6.0 million hectares in 40 *soums*, with the aim to benefit about 20,000 families. Investment activities will include upgrading of livestock water points, pasture management, raising of livestock productivity, improved breeding stock and animal health, small-scale irrigation, fodder production, employment creation, and strengthened marketing systems.

(iii) **Output 3: Capacities and reforms for sustainable livestock management are developed.** Working with the local government and herder groups, the project would provide the institutional support to develop the necessary reforms, policy, and regulatory mechanisms for sustainable pasture and livestock management, including allocation of land rights to improve productivity while reducing stocking numbers. Training and development of capacities and support for capacity building will be provided in the *bag* and *khotail*. Microfinance for herders will also be established to procure inputs and finance for improved livestock production and WSS. Initiatives to promote private sector investment will likewise be supported. Monitoring systems will be set up to evaluate project outputs.

1.4 Outline Investment Plan

The project is estimated to cost $30.0 million, which includes $1.5 million for technical assistance to support planning and design of the project, $25.5 million in investment, and $3.0 million to support capacity building as well as institutional strengthening and reforms. A consulting firm, civil society organization, or pool of individual consultants will be recruited to plan and design project activities, and to provide support for their implementation. Funding for the project is proposed as follows: 60% government, 30% beneficiaries, and 10% private sector. The outline investment plan is shown in Table A1.2.

Table A1.2: Outline Investment Plan

No.	Output	Amount ($ million)
1	Plans for improved rural WSS and sustainable livestock management	1.5
2	Investment in rural WSS, livestock water points, and sustainable pasture and livestock management over an area of 6.0 million ha	25.5
3	Capacity building and reforms for rural WSS and sustainable livestock management	3.0
	Total	**30.0**

ha = hectare, WSS = water supply and sanitation.
Source: Asian Development Bank. 2017. *Country Water Security Assessment of Mongolia*. Consultant's report. Manila (TA 8855-MON).

1.5 Implementation Arrangements

The project duration will be 3 years starting in 2020. The Ministry of Food, Agriculture, and Light Industry will be the executing agency. A project implementation unit will be established in each of the nine *aimags*, and a project steering committee will be established under the Ministry of Food, Agriculture, and Light Industry to ensure close coordination and national ownership.

Appendix 2

Project Concept Notes for Possible Priority Investment Projects: Strategic Investment for Water Supply and Sanitation in *Ger* Areas of Ulaanbaatar

Table A2.1: Project at a Glance

Project Title	Strategic Investment for Water Supply and Sanitation in *Ger* Areas of Ulaanbaatar	
Summary	The project will develop cost-effective and integrated approaches to establish WSS facilities in the *ger* areas of Ulaanbaatar, including improved access to potable water, hygienic disposal of gray water, improved drainage, and development of low-cost sanitation.	
A. General Information		
Area of Water Security	*Investment Sector*	*Supporting Initiatives*
1. Household	■ Herder WSS	☐ IWRM
2. Economic	☐ Irrigation ☐ Livestock ☐ Mining ☐ Industry ☐ Energy	■ Water quality ■ Surface water ■ Groundwater ☐ Agriculture ■ Poverty ■ Private sector PPP
3. Urban Water Supply	■ Ulaanbaatar water supply ☐ Major urban centers ☐ *Soum* centers	
4. Environment	☐ River basin governance ☐ Overall water resources management	☐ Catchment and protected areas ☐ Water resources protection and enhancement
5. Resilience to Water-Related Disasters	☐ Urban flood protection and management ☐ Integrated disaster management	
B. Key Thematic Areas		
■ Sustainable economic growth ■ Inclusive social development ■ Governance ■ Gender and development	☐ Environmental sustainability ☐ Regional cooperation ■ Private sector development ■ Capacity development	

continued on next page

Table A2.1 *continued*

C. Coverage
■ Districts of Ulaanbaatar: Chingektai and Sykhbaatar, Gachuurt, and Ulaisatai ■ River basin: Tuul River Basin
D. Responsible and Supporting Units
■ Responsible ministry: Ministry of Construction and Urban Development ■ Supporting unit: Municipality of Ulaanbaatar
E. Linkage to the Investment Program

1. Technical assistance
  (i) Urban WSS policies and strategies studies
  (ii) Detailed planning and design for Ulaanbaatar urban area
 	
2. Investment
  (i) Ulaanbaatar urban WSS investment program
  (ii) Integrated investment for urban flood protection and management

IWRM = integrated water resources management, PPP = public–private partnership, WSS = water supply and sanitation.
Notes:
1. *Ger* areas refer to Mongolia's traditional tent communities.
2. *Soum* is Mongolia's subprovincial administrative unit.
Source: Asian Development Bank. 2017. *Country Water Security Assessment of Mongolia*. Consultant's report. Manila (TA 8855-MON).

2.1 Project Rationale

A major source of public health risks as well as soil contamination and water pollution comes from the poor water supply and sanitation (WSS), and the improper disposal of human waste and gray water (water from washing) in the peri-urban areas of Ulaanbaatar. In 2014, about 180,000 households (equivalent to 65% of Ulaanbaatar's population) were reported to have no sewage network connections. They relied on using pit latrines (about 310,000) and open wastewater pits (140,000) for their sanitary needs, gray water, and solid waste disposal. These facilities are mostly unlined, poorly constructed, and badly managed. Access to clean water supply is also limited, with very few house connections and most families reliant on restricted access to water from kiosks or open wells (often at some distance from the households). Water supplied by a kiosk typically provides about 10 liters per capita per day, which is less than half the recommended minimum of the World Health Organization.

There are major ongoing programs for Ulaanbaatar to expand the water and sewerage networks. However, there is limited attention to the poorer outlying *ger* areas (Mongolia's traditional tent communities), and many services will not reach them for many years. There is an urgent need for improved levels of service for water supply together with permanent on-site or semipermanent household-level sanitation measures, which are simple, replicable, and cost-effective. Other towns and cities in Mongolia have similar problems, albeit at much smaller scales.

The WSS issues in the peri-urban *ger* areas affect the urban centers in Ulaanbaatar as well as *aimag* and *soum* centers. WSS problems are characterized by unplanned development, inadequate road networks, and a severe lack of social and economic facilities and basic infrastructure and services for water, sewerage, and heating. Outlying *ger* areas are commonly not connected to the centralized networks of water, sewerage, and drainage systems, and unsanitary living conditions are therefore widespread. The project will serve as a model to scale up improved water supply and on-site sanitation for low-income communities in other urban centers, and will complement infrastructure development in central areas.

The overall development of improved WSS in Ulaanbaatar's *ger* areas is constrained by high construction costs, lack of urban planning, and inadequate infrastructure. The lack of basic infrastructure limits economic growth and increases negative environmental impacts. The *ger* areas are predominantly residential, but with pockets of economic growth as well as needs for public and commercial services. The *gers* are gradually being replaced with houses, and the Ulaanbaatar 2020 Master Plan sets out proposals to resettle some of the *ger* residents in apartments.[1]

Current strategies for water supplies to the *ger* areas are limited to standpipes, often at some distance to the households, and households frequently pay higher costs for water than in the central areas. There is very limited consideration for sanitation in the *ger* areas, and most households remain dependent on uncontained pit latrines and wastewater pits. Improved access to safe drinking water has resulted in increased gray water, which is resulting in environmental degradation and potential health risks in areas with no sewerage system. Gray water is discharged untreated into pits, open drainage channels, or on open land.

There are significant socioeconomic variations in *ger* areas as well as rapid changes, with many traditional *gers* being replaced by houses and apartments. Therefore, there is a need for dynamic approaches to meet current as well as future needs.

2.2 Objectives

The project will develop cost-effective and integrated approaches to establishing appropriate WSS facilities in the *ger* areas, including access to clean water from standpipes or house connections, hygienic disposal of gray water, improved drainage, and development of low-cost sanitation. The project will focus on the *ger* areas of Ulaanbaatar in the three districts of Chingektai and Sykhbaatar, Gachuurt, and Ulaisatai. The project is projected to support improved WSS for about 20,000 families.

2.3 Impact, Outcome, and Outputs

The project impact will be improved health and social well-being in the *ger* areas of Ulaanbaatar. The project outcome will be that about 20,000 families living in the three districts have access to improved WSS. The project will have three outputs:

(i) **Output 1: Planning for water supply and sanitation in the peri-urban *ger* areas of Ulaanbaatar is strengthened.** This output will (a) establish working groups of stakeholders to support the planning processes and coordinate initiatives for WSS in the *ger* areas; (b) review and assess different levels of service for WSS, including options for centralized and decentralized systems incorporating lessons learned from previous programs; (c) examine how investments in WSS for the *ger* areas can be designed to support the establishment of *ger* areas as hubs for socioeconomic development of the rural areas; and (d) assess financing and management options, including public–private partnerships, community management, strengthening of self-investment through microfinance, and review of cost recovery options, including prepaid metering.

---

[1]   The Asia Foundation. 2014. *Ulaanbaatar 2020 Master Plan and Development Approaches for 2030.* Ulaanbaatar.

(ii) **Output 2: Construction of improved water supply and on-site sanitation facilities for peri-urban *ger* areas.** This output covers the construction of WSS facilities for about 20,000 families in the *ger* areas in three districts of Ulaanbaatar, including investment in water distribution; house connections; and on-site sanitation for domestic use, small industries, and public and municipal services.

(iii) **Output 3: Systems for sustainable financing and management are developed.** This output is designed to establish financing and management systems to meet the investment and operational requirements of the domestic supply, small industry, and public and municipal services in the *ger* areas. Support will be provided for household connections and improved pit latrines, including giving access to low-cost credit to meet costs of self-connection for water supply and improved pit latrines. The output will also help strengthen policy and strategies to support new financing and management initiatives, including soft public–private partnership or community-based management systems.

2.4 Outline Investment Plan

The project is estimated to cost $21.0 million, which includes $0.5 million for technical assistance to support planning and design of the program, $20.0 million in investment, and $0.5 million to support capacity building and reforms. A consulting firm, civil society organization, or pool of individual consultants will be recruited to plan and design project activities, and to support their implementation. Project funding is proposed as follows: 40% government, 30% beneficiaries, and 30% private sector finance. The outline investment plan is shown in Table A2.2.

Table A2.2: Outline Investment Plan

No.	Output	Amount ($ million)
1	Planning for water supply and sanitation in the peri-urban *ger* areas of Ulaanbaatar	0.5
2	Construction of improved water supply and on-site sanitation facilities for peri-urban *ger* areas	20.0
3	Development of sustainable financing and management of improved water supply and sanitation	0.5
	Total	**21.0**

Note: *Ger* areas refer to Mongolia's traditional tent communities.
Source: Asian Development Bank. 2017. *Country Water Security Assessment of Mongolia*. Consultant's report. Manila (TA 8855-MON).

2.5 Implementation Arrangements

The project duration will be 3 years starting in 2020. The Municipality of Ulaanbaatar will be the executing agency and will establish a project implementation unit to support implementation. A project steering committee will be established under the Ministry of Construction and Urban Development to ensure close coordination and national ownership.

Appendix 3

Project Concept Notes for Possible Priority Investment Projects: Integrated Investment for Flood Protection in Urban Areas

Table A3.1: Project at a Glance

Project Title	Integrated Investment for Flood Protection in Urban Areas	
Summary	The project will support studies and investments in integrated flood protection and management in selected urban locations in three river basins.	
A. General Information		
Area of Water Security	*Investment Sector*	*Supporting Initiatives*
1. Household	☐ Herder WSS	☐ IWRM
2. Economic	☐ Irrigation ☐ Livestock ☐ Mining ☐ Industry ☐ Energy	☐ Water quality ■ Surface water ☐ Groundwater ☐ Agriculture ☐ Poverty ■ Private sector PPP
3. Urban Water Supply	■ Ulaanbaatar water supply ■ Major urban centers ■ *Soum* centers	
4. Environment	☐ River basin governance ☐ Overall water resources management	☐ Catchment and protected areas ☐ Water resources protection and enhancement
5. Resilience to Water-Related Disasters	■ Urban flood protection and management ☐ Integrated disaster management	
B. Key Thematic Areas		

■ Sustainable economic growth ■ Inclusive social development ■ Governance ☐ Gender and development	■ Environmental sustainability ☐ Regional cooperation ☐ Private sector development ☐ Capacity development

continued on next page

Table A3.1 *continued*

C. Coverage
■ River basins: Kharaa, Orkhon, and Tuul
D. Responsible and Supporting Units
■ Responsible ministry: Ministry of Construction and Urban Development ■ Supporting unit: *Aimag* government
E. Linkage to the Investment Program
1. Technical assistance   Integrated planning and design for water-related disasters
2. Investment   Integrated investments for urban flood protection and management

IWRM = integrated water resources management, PPP = public–private partnership, WSS = water supply and sanitation.

Note: *Aimag* refers to the provincial administrative unit in Mongolia, whereas *soum* is the subprovincial administrative unit.

Source: Asian Development Bank. 2017. *Country Water Security Assessment of Mongolia.* Consultant's report. Manila (TA 8855-MON).

3.1 Project Rationale

Mongolia faces a wide range of water-related disaster risks. Urban flooding, one of the main challenges, has been exacerbated in recent years by rapid urbanization and the increased frequency of extreme rainfall events resulting from climate change.

Floods in Mongolia are caused by snowmelt in the spring as well as heavy rains in the summer. Periodic intense rain from June to September is the main cause of flooding, although rapid snowmelt and rain in April can also be problems. Although floods occur in the rural areas, the impacts there are relatively small; the main impacts are in the urban areas.

More frequent extreme weather events under climate change are projected. These events, together with continued urbanization, will increase flood risks and increase the costs of providing sufficient flood control infrastructure. The key challenge to urban flood risk management is determining the most effective means of investing scarce resources to meet potential and uncertain risks. This requires enhanced knowledge, increased awareness, better assessment of risks, and a strong cross-sector institutional framework. Development of new and more integrated approaches to flood management is required.

Physical investments to reduce disaster risk are possible solutions to meet specific needs. However, costs are high and, in many cases, only benefit a limited number of people. The engineering-focused approach does not adequately incorporate the benefits and opportunities of integrated flood management. Many high-profile disasters have shown that communities with engineered protection can face significant and even increased flood risks from events such as embankment failure. An integrated approach can reduce the human and socioeconomic impacts of floods and incorporate social, economic, and ecological benefits in the management of flood plains.

Examples of nonstructural measures that can be mainstreamed to complement structural flood solutions include spatial land use planning, flood plain zoning, provision of areas for relief and flood retention, and mechanisms to improve evacuation of water. Parallel programs need to be considered, including managed aquifer recharge in flood zones; catchment and wetland management; and flood warning, flood insurance, and flood compensation schemes.

Investments for urban flood management need to be tailored to the specific requirements of Mongolia's towns and cities, taking the following factors into account:

(i)   Mongolia is well endowed with land, and the government will, in general, not invest in the protection of flood plains.

(ii)   Urban planning needs to provide an adequate buffer and setback of housing from natural drainage channels, especially in the hill areas, where flash floods are dangerous and almost impossible to manage.

(iii)   More effort should be made to take natural flood retention measures, including maintaining open areas upstream or in the urban areas as flood retention basins. Flood retention can support groundwater recharge, and basins can be used for recreation or parks.

(iv)   Low-cost earth drains should be developed rather than concrete drains, wherever possible.

(v)   Much of the flooding is because of poor drainage maintenance, and ensuring drains are maintained and cleared is one of the lowest-cost solutions.

According to the World Bank's urban flood risk assessment for Ulaanbaatar,[1] the total investment required for implementation of the proposed projects was estimated at MNT1,430 billion ($636 million) at 2016 exchange rates; nearly 87% of the estimated investment was for hard structural measures, and the remaining 13% was for soft structural and nonstructural measures. The Ulaanbaatar 2020 Master Plan identifies a slightly different approach, including three options: (i) enforcing and relocating persons away from the flood plains, including resettlement ($200 million); (ii) rehabilitating dykes through hard investments in storm drains ($60 million); and (iii) supporting community-based management, including awareness raising ($3 million).[2] The master plan also identifies programs for forest protection and management ($4 million).

There are no definitive estimates of costs of urban flood protection outside Ulaanbaatar. The investment plan assesses the costs of urban flood protection, including Ulaanbaatar, *soum,* and *aimag* centers, to be $280 million up to 2030. This is based on lower-cost integrated approaches that would be more likely to be within the available budget of the government.

3.2 Project Objectives

The objectives of the project are to (i) review the key risks of urban flooding and identify the most feasible areas of investment targeted at increasing security and reducing risks from water-related disasters in three river basins, (ii) support investment in integrated approaches for selected components, and (iii) strengthen the mechanisms and institutions to support integrated flood management.

---

[1]   World Bank. 2015. *Flood Risk Assessment and Management Strategy of Ulaanbaatar City.* A joint study by World Bank's Agriculture Risk Study Center, JEMR, and Usny Erchim. Ulaanbaatar.

[2]   The Asia Foundation. 2014. *Ulaanbaatar 2020 Master Plan and Development Approaches for 2030.* Ulaanbaatar.

3.3 Impact, Outcome, and Outputs

The project impact will be increased protection and resilience in urban flood-prone areas in Mongolia. The outcome will be improved knowledge for decision-making and investment in integrated urban flood protection and management in three river basins. The project will have three outputs:

(i) **Output 1: Integrated flood management plans and designs are prepared.** This output includes the following: (a) studies to identify priorities and cost-effective solutions to urban flooding, incorporating assessment of climate change impacts and adaptation measures; (b) analysis of economic benefits and financial returns of soft and hard strategies for flood protection and management; and (c) preparation of flood management plans for three river basins, and feasibility studies and detailed designs for priority projects.

(ii) **Output 2: Urban economies are improved through investment in integrated flood management.** The project will support investments in flood protection and management in selected urban locations in three river basins: Kharaa, Orkhon, and Tuul. The project will be directed at integrated approaches applying low-cost and sustainable initiatives, incorporating improved drainage, protection, land use planning, awareness, and maintenance of flood infrastructure.

(iii) **Output 3: Integrated flood management is strengthened.** The project will strengthen the mechanisms and institutions for integrated flood management, including the mainstreaming of flood protection strategies into sector programs, institutional strengthening and reforms, engagement with communities, and training and awareness. This output will also aim to strengthen the role of river basin organizations and the Ministry of Environment and Tourism to provide improved knowledge on climate and weather patterns under climate change, and guidance on how these can be applied to urban flood management.

3.4 Outline Investment Plan

The project is estimated to cost $22.0 million, which includes $1.5 million for technical assistance to support planning and design of the project, $20.0 million in investment, and $0.5 million to support institutional strengthening and capacity building. A consulting firm, civil society organization, or pool of individual consultants will be recruited to plan and design project activities, and to support their implementation. Project funding is proposed as follows: 80% government, 10% beneficiaries, and 10% private sector. The outline investment plan is shown in Table A3.2.

Table A3.2: Outline Investment Plan

No.	Output	Amount ($ million)
1	Planning and design	1.5
2	Investment in integrated flood management in three river basins	20.0
3	Strengthened integrated flood management	0.5
	Total	**22.0**

Source: Asian Development Bank. 2017. *Country Water Security Assessment of Mongolia.* Consultant's report. Manila (TA 8855-MON).

3.5 Implementation Arrangements

The project duration will be 3 years starting in January 2020. The Ministry of Construction and Urban Development will be the executing agency. A project implementation unit will be established in each of the three river basins, and a project steering committee will be established under the Ministry of Construction and Urban Development to ensure close coordination and national ownership.

www.ingramcontent.com/pod-product-compliance
Lightning Source LLC
Chambersburg PA
CBHW051657210326
41518CB00026B/2620